PHILOPONUS

On Aristotle

Physics 4.1-5

PHILOPONUS

On Aristotle Physics 4.1-5

Translated by Keimpe Algra
and
Johannes van Ophuijsen

B L O O M S B U R Y
LONDON • NEW DELHI • NEW YORK • SYDNEY

Bloomsbury Academic
An imprint of Bloomsbury Publishing Plc

50 Bedford Square
London
WC1B 3DP
UK

1385 Broadway
New York
NY 10018
USA

www.bloomsbury.com

Bloomsbury is a registered trade mark of Bloomsbury Publishing Plc

First published in 2012 by Bristol Classical Press
Paperback edition first published 2014

British Library Cataloguing-in-Publication Data
A catalogue record for this book is available from the British Library.

ISBN HB: 978-1-7809-3211-8
PB: 978-1-4725-5800-8
ePDF: 978-1-4725-0177-6

Library of Congress Cataloging-in-Publication Data
A catalog record for this book is available from the Library of Congress.

The present translations have been made possible by generous and imaginative funding from the following sources: the National Endowment for the Humanities, Division of Research Programs, an independent federal agency of the USA; the Leverhulme Trust; the British Academy; the Jowett Copyright Trustees; the Royal Society (UK); Centro Internazionale A. Beltrame di Storia dello Spazio e del Tempo (Padua); Mario Mignucci; Liverpool University; the Leventis Foundation; the Arts and Humanities Research Council; Gresham College; the Esmée Fairbairn Charitable Trust; the Henry Brown Trust; Mr and Mrs N. Egon; the Netherlands Organisation for Scientific Research (NWO/GW); the Ashdown Trust; Dr Victoria Solomonides, the Cultural Attaché of the Greek Embassy in London. The editor wishes to thank Gábor Betegh, Barrie Fleet, Michael Griffin, Pamela Huby, Richard McKirahan, and Arnis Ritups for their comments, Ian Crystal for preparing the volume for press, and Deborah Blake, who has been the publisher responsible for every volume since the first.

Typeset by Ray Davies

Contents

Conventions

<...> proposed addition to the Greek text.
[...] proposed deletion from the Greek text.
(...) parentheses inserted whether by the editor or by the translators.
{...} to be supplied in thought.

Bold type is used for words quoted by Philoponus from Aristotle's *Physics* text.

Introduction

Keimpe Algra

Philoponus' commentary on Aristotle's discussion of place in *Physics* 4.1-5 is nowadays usually treated as consisting of two distinct parts: the commentary proper on the text, in nine sections or lectures, and the so-called 'corollary on place' which has been inserted as an excursus at the end of the seventh lecture. It is fair to say that the commentary proper has received far less attention from modern scholars than the corollary. This may be partly due to the fact that the corollary was translated separately into English as one of the first volumes in this Ancient Commentators on Aristotle series, whereas the English translation of the rest of the commentary on *Physics* 4 is appearing in three volumes only in 2012.[1] Apart from this, however, it is also hard to deny that the running commentary on the one hand, and the corollary on the other, are not on an equal footing in terms of philosophical power and significance. The corollary contains a fairly systematic attack on Aristotle's conception of place, which clearly uses earlier discussions of problems in Aristotle's theory (thus, four out of the five *aporiai* put forward by Theophrastus (fr. 146 FHSG) reappear, although without ascription) and an equally systematic exposition and defence of an alternative conception of place as a three-dimensional extension.[2] The running commentary, on the other hand, is predominantly exegetical and paraphrastic, stays close to Aristotle's text and in general, though certainly not always, refrains from taking an independent critical stance. It may be a fine example of a late antique scholastic commentary, and it does contain exegetical exercises which do have a philosophical interest in their own right, such as the attempt to offer a coherent explanation of the way in which the terms 'in respect of itself' (*kath' heauto*), 'in respect of something else' (*kat' allo*), 'in the primary sense' (*prôtôs*) and 'incidentally' (*kata sumbebêkos*) should be used (*in Phys.* 530,1-531,5). Yet we do not see too much here of the dragon slayer of Aristotelian physics who Philoponus is often taken to be. Here we may have another reason why, in modern scholarship, the commentary proper has been overshadowed by the corollary.

The nine sections into which the commentary proper on *Physics*

4.1-5 is divided each have the same form – a form which seems to reflect the classroom practice of similarly structured *praxeis* or lectures, of which we also find traces in the works of such later Neoplatonists as Olympiodorus, Elias and David. Each chapter starts out with what Philoponus himself labels the *theôria*: a kind of sustained paraphrasing and explanatory lecture on the text at issue, which is then followed by a commentary on the *lexis*, i.e. a commentary on the wording of individual passages, keyed to lemmata taken from Aristotle's text.

Philosophically the *theôriai* are the most interesting sections, but even these are on the whole didactic rather than scholarly in nature. Philoponus' primary intention seems to be to lay bare the structure of Aristotle's thought by showing how the various topics broached in individual sections of *Physics* 4.1-5 hang together, by formalising the argument in terms of categorical or hypothetical syllogisms, by paraphrasing and by explaining the vocabulary chosen by Aristotle. In this respect Philoponus' commentary on the *Physics* differs considerably from the one written more than a decade later by Simplicius. It is hard to imagine Simplicius' huge commentary functioning in an everyday classroom practice. Simplicius offers a host of relevant quotations from earlier philosophers and his commentary is replete with references to interpretations provided by the earlier commentary tradition. In general (i.e. apart from the corollaries) Philoponus' text is more elementary, and in some respects it comes closer to the paraphrasing exegesis of Themistius, by which it has clearly been influenced. Philoponus often refers to Themistius, he sometimes takes over his readings of Aristotle's text, and he includes extensive paraphrasing quotations from his work, e.g. in his discussion of Aristotle's rather obscure arguments against the conception of place as a three-dimensional extension (*in Phys.* 550,9-551,20). In addition, Philoponus apparently knew and used the now lost commentary on the *Physics* by Alexander of Aphrodisias. It is to the latter that he seems to owe his typical way of highlighting the structure of the argument in Aristotle ('having established that X he now goes on to show that Y, etc.').

The way in which the commentary proper and the corollary relate deserves some closer examination. It has been argued by Koenraad Verrycken that a comparison of Philoponus' works reveals a development in his metaphysical stance from an early stage of Alexandrian Neoplatonism, which Verrycken labelled 'Philoponus 1', to a plain Christian metaphysics involving, among other things, the rejection of the eternity of the world.[3] According to Verrycken, the *Physics* commentary as we have it combines traces of both stages of this development and should be assumed to consist of a substrate representing Philoponus 1, which was written in 517,[4] plus a number of additions postdating the year 529, when Philoponus published his polemical *On the Eternity of the World against Proclus*. According to Verrycken, the corollary on place should be also taken to belong to the later additions, because it defends the conception of place as an independent three-dimensional extension, whereas the main text of the commentary appears to endorse and even to defend Aristotle's conception of place as the immobile surface of the surrounding body. This is not the place to deal with Verrycken's view of the chronology of Philoponus' works in general, which has proved quite influential in the meantime. Let us focus instead on the arguments that concern the chronological and systematic relation between the commentary proper on *Physics* 4.1-5 and the corollary on place.

As a preliminary point it is worth noting that, unlike the rejection of the eternity of motion and of the world, the new conception of place advocated in the corollary does not appear to have any obvious connection with the supposed Christian metaphysics of Philoponus 2. So even if we could be sure that it constitutes an addition (which we cannot), and even if we could be sure that the Christian metaphysics of Philoponus 2 postdates 529 (which we cannot either), there is no reason to assume that the corollary on place must have been written as late as 529, or even later. But there is more. Doubts can be raised both concerning Philoponus' supposed commitment to Aristotle's position in his commentary and concerning the claim that his endorsement of the conception of place as a three-dimensional extension represents a later development in his thought.

First, as already indicated, the general purpose of the commentary appears to be to set out and paraphrase Aristotle's text in a way that makes it maximally comprehensible and coherent. Accordingly, the commentator for the most part – though not exclusively, as we shall see – takes on the *persona* of Aristotle, so to speak, and his subject matter is not so much the world according to Philoponus, but the world as conceptualised by Aristotle in the particular passage he is each time commenting on. Indeed, in the introduction to Philoponus' commentary on the *Categories* this practice of keeping commentary and critical evaluation apart is even explicitly advocated as the right way to proceed for any commentator:

> [the commentator] has to be a scrupulous judge of everything that is said and he must first explain the meaning of the ancient text and interpret the doctrines of Aristotle and *then* express his personal opinion (*in Cat.* 6,30-5).

It is in accordance with this preferred practice that the corollaries on place and void, which both take a broader perspective, are inserted as excursus. Indeed, at the end of his second *theôria* on *Phys.* 4.4 Philoponus announces that he will add what we now know as the corollary on place after having furnished the obligatory comments on the *lexis*, and the wording of this announcement clearly indicates the shift of perspective:

> These are the attempts culled from the exegetical tradition devoted to the Aristotelian text that are intended to establish that place is not an extension. The external arguments which the commentators have added, and whatever the proponents of the view that place is an extension could say, we will expound after having gone through the text (*in Phys.* 552,10-14).

In other words: thus far the commentary has followed the text and the Aristotelian exegetical tradition, so it is not surprising that it has defended Aristotle's view; the corollary, by contrast, will weigh arguments that are not necessarily related to Aristotle's text ('external arguments'), and this means that there Philoponus may come to a different conclusion. To this extent the *Nebeneinander* of the two perspectives is perfectly intelligible from the different natures and purposes of the commentary proper and the corollary respectively, and we need not have recourse to the hypothesis of two strata.[5]

As for the passages which have been supposed to betray Philoponus' own endorsement of Aristotle's position, it has recently been shown by Pantelis Golitsis that they are almost invariably simple *calques* of what is in Aristotle's text, and that the use of the first-person singular verbal forms *phêmi* or *legô* in Philoponus' commentary does not signal his commitment to Aristotle's position either. Indeed, these terms turn out to have explicative rather than assertoric force, carrying the connotation 'I mean' or 'that is', rather than 'I claim'.[6] If this shows that there are no compelling reasons to assume that Philoponus positively endorsed Aristotle's position at the time of writing the commentary, we may go one step further and observe that, on the contrary, there are some positive indications that at that time Philoponus already endorsed his own conception of place as a three-dimensional extension. First of all, we may note the different ways in which the commentary addresses the arguments which Aristotle raises against the various rival views. It claims that the Aristotelian arguments against the identification of place with matter or form are 'compelling' (*anagkaioi*; *in Phys.* 548,14), whereas it introduces Aristotle's argument against the conception of place as extension by claiming that 'the meaning of his words is very unclear', while adding that 'different interpreters try to grasp the meaning of what he says in different ways' (*in Phys.* 548,16-18).

Secondly, there are some passages in the commentary which seem to betray that Philoponus had his own conception of place as a three-dimensional extension at the back of his mind all along. In so far as these passages definitely go beyond what Aristotle can have meant, they may even be regarded as slips of the commentator *qua* commentator. Thus, Philoponus explains one of the basic characteristics of place listed by Aristotle, viz. that 'every place has "above" and "below" ' (211a3-4) by saying that this does not refer 'simply to each individual place (for each individual place is *either* above *or* below); but in place as a whole there is above and below' (*in Phys.* 541,6-7). However, on Aristotle's conception of place as a surrounding surface, there is no such thing as 'place as a whole', whereas on Philoponus' own conception of place as extension such a phrase becomes intelligible. Next, in his commentary on chapter 5 (*in Phys.* 597,32-598,2), in the context of his exegesis of Aristotle's claim that his own conception of place as a surrounding surface allows us to solve the puzzle that when a body grows place may have to grow along with it, Philoponus uses the expression 'takes up place' in the un-Aristotelian sense of 'occupies space': 'it is clear that the body that has grown takes up just that place which was occupied by the food' (*in Phys.* 598,1-2). And in the course of his comments on Aristotle's account of the natural motions of the elements in chapter 5 of *Physics* 4, Philoponus suddenly introduces the un-Aristotelian notion of 'the force of the void' (*in Phys.* 600,6). Precisely because these passages do not signal that Philoponus is speaking *in propria persona*, but occur as part of his exegesis of Aristotle's text, which they fit badly, they are unlikely to be consciously made later additions. They rather read like indications that despite his attempts to offer a truthful account of Aristotle's position he couldn't help talking in terms of his own preferred conception from time to time.

Thirdly, and more importantly, the part of the commentary which follows on the corollary contains some *explicit* references to Philoponus' preferred conception as one which is at least as adequate as its Aristotelian counterpart. Let me just quote one of them:

> The following as well was a characteristic of place: that it is neither larger than that which is in place nor smaller, and this as well is part of this account of place. For if it is the limit of what contains, and the limits of the container and of what is contained are together (for the surfaces fit onto each other), then the place is neither larger nor smaller, but equal in size. Equal in size of course with respect to the circumference – for it is not equal with respect to the whole three-dimensional extension. In consequence those who say that place is an extension, may have a more reasonable way of saving the idea that place is equal in size to the thing that is in it. For on this view it will be equal in every dimension (*in Phys.* 587,22-30).

Is Philoponus here departing from his professed policy of keeping explanation and critique apart? Not really, it seems. For one thing, these passages are not as explicitly critical of Aristotle's theory itself as is the corollary. They merely present the conception of place as extension as an alternative conception which is at least equivalent in the respects highlighted by Aristotle. For another thing, in the context at issue, which is the commentary on the last part of chapter 4 and on chapter 5 of *Physics* book 4, this procedure makes sense, since it is only there that Aristotle tries to show that his own conception, and by implication *not* the rival conception of place as extension endorsed by Philoponus, best tallies with the common conceptions concerning place, and that it alone allows us to solve all the *aporiai* that have been mentioned. In other words, Aristotle's claim that his own conception of place is in these respects superior involves an implicit comparison between it and the most important rival conception. Accordingly his commentator may feel entitled to offer such a comparison as well.

Finally, the corollary itself also contains an indication that Philoponus had already developed his own ideas on place at a relatively early date. For it refers to a defence of Aristotle's position against Philoponus' objections by his teacher Ammonius, 'the Philosopher':

> However, when we made these points against what Aristotle said about place, a defence was put forward by the Philosopher (*in Phys.* 583,13-15).

These words suggest that Philoponus had already made his critical observations during Ammonius' classes, or at any rate at a time when Ammonius was still around and philosophically active. As Verrycken himself indicates, we do not know whether Ammonius was still active or even alive at the time when the supposed Philoponus 1 wrote what Verrycken takes to be the first version of the *Physics* commentary, i.e. in 517.[7] According to Westerink's chronological reconstruction, Ammonius was probably born between 435 and 445, so if alive at all, he should have been between 72 and 82 years old in the year 517.[8] It is unlikely, at any rate, that he would have been active for much longer, perhaps even more unlikely that he would have been still alive when Philoponus 2 supposedly added the critical corollary to his *Physics* commentary after 529, at least twelve years later.

All in all then, the evidence seems to suggest that Philoponus' critique of the Aristotelian theory of place originated earlier in his career and that the juxtaposition of passages that explain and defend Aristotle on the one hand, and passages that are critical of his position on the other, is to be ascribed to Philoponus' conception of the duties of a commentator on a text like this. If the commentary and the corollary can thus be regarded as two aspects (exegetical *versus* critical) of the same project, they can be used to explain each other. Two examples, both concerning interesting philosophical questions to do with place, may serve to illustrate this. The first one concerns Philoponus' interpretation of Aristotle's position on the causal status of place, the second his own view on the emplacement of parts of continuous substances.

Aristotle's statements about the role of place in the explanation of natural motion at first sight seem to point in different directions: most importantly, place does appear to have a certain *dunamis* or power (208b10), yet it is not one of the four causes (209a20). I have argued elsewhere that giving due attention to the dialectical structure of the discussion in *Physics* 4 and to the additional evidence in *De Caelo* and *Physics* 8 should lead us to conclude that for Aristotle place is not itself a cause, final or otherwise, of natural motion. It is rather the element's *being in* a natural place which, as a concomitant to the actualisation of its form, is the final cause of its natural motion. The role of the concept of place in the explanation of natural motion then seems to be that it allows us to specify this 'being somewhere' which constitutes the goal of the natural motion of the elements.[9] This interpretation thus does allow a role for place in the explanation of natural motion, but not as a cause. However, we can for the moment ignore the question what Aristotle himself really thought, and concentrate instead on how Philoponus interpreted his position.

It appears that, on this subject, Philoponus did not have much to go on in the earlier commentary tradition. Simplicius, at any rate, tells us that the question what causal status Aristotle was willing to accord to place is a matter which still requires further examination, 'because the commentators pass it over in an off-hand fashion' (*in Phys.* 533,26-30). This claim seems to be confirmed by the fact that Themistius' paraphrase devotes only a few lines to the subject:

> For how will it {i.e. place} be a first principle? As matter is? So what is compounded from it? Or is it like form or what causes change? Or like that for the sake of which? Yet in what way?' (*in Phys.* 105,9-11).

These questions are no doubt rhetorical, for they are meant to paraphrase Aristotle's claim that place is not one of the four familiar causes (209a20). Yet, as we saw, the problem is that it is *prima facie* unclear how all this relates to the claim that place has a certain power (*dunamis*, 208b10). Philoponus' commentary deals with this problem in the following way. In his first lecture he simply translates the apparent characteristic about the power (*dunamis*) of place in terms of its acting as a final cause. Nevertheless, in the second lecture, he makes it quite clear that he believes that this is only what place *appears* to be and that it is Aristotle's considered view that place is *not* a cause. He defends this view by comparing what we might call the semantics of final causation and the semantics of natural motion:

> Nor as a final cause (which is actually surprising; because place appears to be like an end or an object of striving). After all, that which is striving for something wants to become that which it is striving for – for example what is striving for the good wants to become good, and what is striving for health wants to become healthy. But none of the things that are in place actually becomes place. So place is not a final cause either (*in Phys.* 509,8-12).

And he adds the following argument:

> Also otherwise: final causes are seen to be present in the things of which they are the ends, but place is different from all the things that are in it, having no share in the emplaced object (*in Phys.* 509,30-510,2).[10]

These passages from the commentary proper may help to put a well-known passage from the corollary into perspective. There Philoponus claims that

> it is quite ridiculous to say that place has any power in its own right: it is not through desire of a surface that things desire that station in the order which they have been given by the Creator (*in Phys.* 581,17-21).

At first sight, and read in isolation, this might seem to be a jab at Aristotle, suggesting that Aristotle did in fact see place as a final cause of natural motion, and as such at odds with the interpretation of the second lecture of the commentary proper. On closer view, however, this does not appear to be the case.

First of all, the passage does not figure in the section of the corollary which specifically criticises Aristotle. That section appears to concentrate exclusively on problems to do with the *morphology* of Aristotelian places as surfaces, and does not discuss the supposed causal status of place. Instead, the passage we are dealing with is to be found in the part of the corollary in which Philoponus defends his *own* conception of place as extension against possible objections. One of these objections, then, is that *if* place should be supposed to have some power, one cannot see how Philoponus' void-space could play that role (579,23-560,3). Furthermore, it is an objection which is introduced with the words 'the following objection might reasonably be raised by someone on behalf of the Aristotelian view'. Such an anonymous 'someone' would take Aristotle's claim about the *dunamis* of place at face value, something which the second lecture of Philoponus' commentary had suggested we should not do. Philoponus now turns the tables on such an opponent by claiming that Aristotle's place-as-a-surface cannot exercise the required force either. There is no suggestion that he is straightforwardly attacking Aristotle himself here.[11] All this suggests that the claim that Philoponus offers a 'complete repudiation of the Aristotelian [...] ascription of power to places' should be qualified.[12] He does indeed repudiate such an ascription of power to places, but he does not repudiate it as a position that was in fact endorsed by Aristotle.

In this case it can be seen that the commentary proper can illuminate the corollary. A second example may serve to show that in its turn the corollary can illuminate claims made in the commentary. At *Physics* 4, 212b22-8 Aristotle claims that on his conception of place the *aporiai* that he has listed earlier can be solved. One of the examples he mentions is that on his view there will be no place of a point (neither of a surface, we may presume). In the part of his commentary which follows on the corollary, and in which, as we saw, Philoponus feels free to compare Aristotle's conception with his own favoured one, he comments on this passage by claiming that on his conception of place as a three-dimensional extension, points and surfaces are not in a place either:

> It is plain, however, that those who claim that place is three-dimensionally extended will say these same things. [...] But neither is it necessary {on this view} that there is a place of a surface, nor of a mark, or point. For if what is in place in respect of itself is the body, what necessity is there for the limits of this body to be in a place {as well}? For if we have shown that not even its parts are necessarily in a place in respect of themselves, how much more does this hold for its limits?' (*in Phys.* 598,3-30).

The last sentence refers back ('it has been shown') to the corollary. There we find the following argument:

> For this reason also the part will not be in itself be in place; for if the body that passed inside the place were actually divided by the extension, it would necessarily follow that because each part is individually outlined each part is individually in place; but if the body is not divided by the extension and the extension does not pass through the body, why should it necessarily follow that the part is in place itself? (*in Phys.* 577,32-580,4).

Individual places, in other words, derive their boundary, and hence their individuality, from the boundary of the occupying substances.[13] The parts of a continuum, however, do not have actual boundaries, hence they are not in individual places in their own right. Of course the claim that such parts *must* have places of their own was something which Aristotle, followed by Themistius, saw as a logical consequence of adopting the conception of place as extension. As Themistius (quoted by Philoponus *in Phys.* 550,30-551,2) put it:

> For if place is an extension, then necessarily each of its parts too will have to be in its own place in respect of itself. After all, each of the parts of the body will occupy a part of its extension of place.

It was indeed this inference which was at the basis of Aristotle's unsatisfactory refutation of the conception of place as extension, which in fact triggered Philoponus' corollary. We can now see that Philoponus refused to go along with it. It seems as if for him, just as for Aristotle, the parts of a continuous substance only have a place incidentally, i.e. their place is the place of the whole of which they are a part. At any rate, he appears to follow Aristotle in his conception of the kind of things that can be said to be in a place in their own right. Also for Philoponus places, properly speaking, seem to be the places of bounded self-subsistent substances. For all his critique and innovation, this is an element of Aristotle's theory which he preferred to leave intact.

These examples may serve to show that, apart from the fact that there are no compelling reasons to assume that the corollary is a later addition reflecting a change of position on Philoponus' part, the commentary proper and the corollary, as parts of the same project, dovetail into each other in several ways and should be studied together, not in isolation.

In our notes to the present translation we signal a number of parallels between Simplicius and Philoponus in their critique both of Aristotle's own conception of place and of his arguments against the rival conception of place as extension. We can see that these common features have no counterpart in Themistius, and they are unlikely to derive from Alexander, who after all was an orthodox Aristotelian. In principle they may be due to the fact that both Philoponus and Simplicius were strongly influenced by their common teacher Ammonius. However, as indicated, the last part of the corollary on place actually suggests that Ammonius *defended* Aristotle against Philoponus' critique (*in Phys.* 583,13-29). Now the way in which this defence is represented there – Ammonius seems to have claimed that Aristotle wanted to come up with a *physical* theory, that a *physical* theory should refer to *physical* entities, and that an immobile self-subsistent extension does not qualify as such – leaves some room for criticisms of what Aristotle *did* in fact write. So we cannot entirely exclude the possibility that Ammonius was indeed the common source in these cases. In addition, it is sound to realise that both Philoponus and Simplicius were indebted to a scholastic tradition (both written and oral) of which we can oversee only a part. Yet, we should perhaps also consider the possibility that Simplicius, when working on this section of the *Physics*, used Philoponus' commentary. True, he seems to have loathed Philoponus as a philosopher in general, but we should also note that his almost hysterical critique of Philoponus as a know-nothing who attacks the great Aristotle in order to become famous is primarily geared to Philoponus' rejection of the theory of the aether and the divine status of the heavenly bodies.[14] This rejection of course threatened Simplicius' typically Neoplatonic psychagogical conception of Aristotle's system, viz. as leading the mind from the sublunary world, through the heavenly stars to the Demiurge. The 'atheist' Philoponus, in his view, was ready to equal the divine light of the stars to the light emitted by worms, and prepared, by denying the eternity of the world, to project his own human finitude on the divine cosmos as a whole. The subject matter of *Physics* 4.1-5, by contrast, does not involve such sensitive issues. In his commentary on this part of the text Simplicius does not criticise Philoponus, as he does elsewhere in his *Physics* commentary, and it does not seem impossible that he used him here without acknowledging it. A further study of the parallels between Philoponus and Simplicius in their commentaries on the *Physics* might be rewarding.

Finally, some remarks on the translation and the footnotes. The translation we offer is based on Vitelli's edition in two volumes of 1887-8 (*CAG* vols 16 and 17; the text of the commentary on *Phys.* 4 is in vol. 17). We have not consulted any manuscripts ourselves, nor have we studied the manuscript tradition afresh, but we offer a list of those cases where we have opted for different readings or ventured conjectures and emendations. The more significant cases are defended in footnotes to the relevant passages.

In order to facilitate an overview of the argument we have briefly set out the structure of each of the nine sections, as well as some aspects of the way in which it relates to Aristotle's text, in a footnote at the beginning of the section at issue.

In those cases where we have inserted translated quotations from the commentaries of Themistius and Simplicius on *Physics* 4, we have used (sometimes with slight changes) the translations of Todd (2003) and Urmson (1992a) and (1992b). For translated quotations from Philoponus' corollary on place we have gratefully used Furley and Wildberg (1991). Our translations from Aristotle are indebted to Hussey (1983) and Waterfield and Bostock (1996).

Acknowledgements

The translators wish to thank the series editor, Richard Sorabji for his trust, and the anonymous readers he marshalled for their constructive suggestions. In addition they are grateful to Ian Crystal and

Sebastian Gertz for their logistical support and for the way in which they have prepared the manuscript for the press. Viivi Lähteenoja and Maarten van Houte offered useful assistance in the process of preparing the manuscript. Finally, Keimpe Algra's contribution to the translation and the notes was facilitated by a research stay at the Fondation Hardt in Vandoeuvres, funded by the Faculty of Humanities of Utrecht University.

Notes

1. The translation of the corollary can be found in Furley and Wildberg (1991). The remaining translations besides the present one are by Sarah Broadie and Pamela Huby.

2. On Philoponus' conception of place as a three-dimensional extension, in itself void, but always filled by bodies, see Sedley (1987).

3. See Verrycken (1990a).

4. At *in Phys*. 703,16-19 Philoponus refers to the current year as the 233rd year of the era starting from Diocletian's reign, which is 517 AD.

5. One may compare the way in which the commentary on book 3 combines purely Aristotelian references to the unchanging aether (*in Phys* 340,31ff.) and to the unmoved mover (*in Phys*. 377,26) in the commentary proper with Philoponus' own exposition on time having necessarily had a beginning (*in Phys*. 456,17-485,30). According to Golitsis (2008) 127-95 there are several smaller corollaries of this kind scattered over the *in Phys*.

6. See in general Golitsis (2008) 27-37; and 34 n. 105 on *phêmi* and *legô*.

7. Verrycken (1990a) 239.

8. Westerink (1962) x.

9. Algra (1995) 195-222, esp. 218-19; comparable views in Machamer (1976) 92 and Morison (2002) 53 (although the latter presents his view as an alternative). *Per contra* Wolf (1987) 96 n. 44 and Sorabji (1988) 186-7 maintain that for Aristotle place is a final cause of natural motion; a more sceptical view in Sorabji (2004) 329.

10. The point that place certainly *appears* to be an object of striving, but on closer analysis is not, is also made by Simplicius (*in Phys*. 533,19-21), who also stresses the fact that, unlike final causes, places are external and self-subsistent (*in Phys*. 533,26-30).

11. Indeed, we may note that the *other* counter-arguments Philoponus is tackling in this part of the corollary (e.g. that he is introducing a self-subsistent quantity, or that on his view space should be infinite rather than finite) are not by Aristotle himself either.

12. The quotation is from Sorabji (1988) 211.

13. We may perhaps compare Philoponus' use, in the corollary on place, of the labels 'up' and 'down' as properties that can be ascribed to space, but that are derived from the bodies occupying it: 'if I must give a general definition, I assert that the part of the extension which receives light bodies is up and that part which receives heavy ones is down' (*in Phys*. 581,33-582,1).

14. See the excellent overview of Simplicius' polemics against Philoponus in Hoffman (1987).

Textual Questions

Deviations from the manuscripts and/or
from the edition by Vitelli in *CAG*.

503,15 *mallon* added after *dêpou*.

506,15 inserting *ou* with Vitelli.

506,17 *holon kath' holon heauto(n)*: perhaps *heautou* should be read instead of the *heauto* and *heauton* of the MSS.

536,1-2 after *to pan to on pou einai*, add *all'oun to en tini einai* (or perhaps *all'oun alêthes to pan to on en tini einai*, or something in between); so in brief, *all'oun* (*alêthes*) *to* (((*pan*) *to*) *on*) *en tini einai*), transposing the adjunct *kata ti tôn tou en tini sêmainomenôn* so as to follow this and precede *pan gar to phusikon anankê en tini einai kai mê en topôi*.

586,2 For *apodexamenês* read *apodexamenon*.

599.25 *ê* added after *êi*.

PHILOPONUS

On Aristotle

Physics 4.1-5

Translation

John Philoponus, Commentary on Aristotle's *Physics* Book 4, Chapters 1 to 5

Chapter 1

208a27 The student of nature must have knowledge about place too, just as he must about the infinite.

The task the writer sets himself in this book is to analyse place, time 496,5
and void, and first of all place.[1] He uses the same method once again. For he first shows that the problem of place is a necessary one to tackle for the student of nature, just as he did in his discussion of the infinite and in fact almost everywhere.[2] That the theory of place is something the student of nature will have to deal with, he establishes on the basis of the opinion of most ancient thinkers. For, he says, they 10
all suppose that the things that exist are in a place, and they establish this by reference to the things that do not exist. For, they argue, if that which is non-existent is nowhere, that which exists must be somewhere, since there cannot exist anything that is not in a place.[3] That they thus convert the argument incorrectly is clear from what has often been remarked.[4] For they convert the argument 15
on the basis of a negation of the antecedent, whereas this should be done on the basis of a negation of the consequent, thus: 'if what is non-existent is nowhere, then what is somewhere must be existent'.[5] In this form the argument would be sound; 'somewhere' is obviously opposed to 'nowhere', and 'existent' to 'non-existent'. Wrong though these people may be in supposing, without further qualification, that 497,1
all the things that exist are in a place, it is at least obvious that the natural bodies, which constitute the subject of the natural philosopher's theory, are in a place. For each of the elements – earth, fire, and the rest – has its well-defined natural place.[6] Therefore it is necessary for the student of nature to have a theory of place.

This is made clear in another way too: if it befalls to the student of nature to analyse change, because he analyses nature, and nature is 5
the principle of change and rest, and if the first and principal type of change is locomotion (for it is shown in the eighth book of this work that none of the other types of change can occur without this one, whereas this one can occur independently of the others: the heavenly bodies, at any rate, exhibit only this type of change), then the student of nature will necessarily have to acquire knowledge of place, which is what this primary type of change relates to.[7] For if the objective of 10
the student of nature is the understanding of natural bodies, and if

each of the elementary bodies has its own natural locomotion, then it is obviously impossible to understand the nature of the simple bodies without understanding their natural motions; and it is impossible to understand their natural motions without knowing the natural place to which each of them moves. Hence, if we do not understand place, we shall not understand the natural motion of the simple bodies, and if we do not understand that, we shall not understand the nature of the simple bodies.[8] And if the simple bodies are not understood, neither will those objects be understood that are composed of them. So we may conclude that ignorance concerning what place is will make it impossible to understand physical objects.

That the theory of place is required for the student of nature is clear from these considerations. But the theory of place is difficult, he says, and puzzling, first of all because when we examine the properties of place, not all these properties point to a single nature for it. Whenever we are making our assertions about things and giving our definitions of them, we should do this on the basis of their natural inherent properties – for example, when we are dealing with man, on the basis of the fact that he is rational, mortal, etc. – but when we are dealing with place, its intrinsic properties suggest to us divergent judgments about it.

For example, it is a property of place to circumscribe and mark off the things that are in it, as a jar does with wine. On these grounds one might suppose that place is form. For it is proper to form to delimit and mark off the things in which it inheres. On the other hand, it is also a property of place to receive ever-different bodies, while remaining one and the same place, as with a jar. For the same vessel can receive wine, and water, and countless other things. To this extent place would seem to be matter, not form. For matter, while remaining one and the same, admits of ever different forms at different times, just as bronze receives at one time the form of a man, at another time the form of a horse. For these reasons the theory of place is a matter of puzzlement, but also, he adds, because we do not possess any **preliminary discussion of puzzles with solutions** about place from earlier thinkers. For no one has adequately discussed the difficulties that are connected with it, so as to bring us even close to what we are looking for (for truly, the starting point for a good solution is puzzlement).[9] Neither has anyone come up with a stock of positive suggestions about place and set them out in a convincing and well-argued way. On the contrary, those who have so much as pronounced on it required that what little they put forward should be believed without any convincing argument. Some of them – Democritus, Leucippus, Metrodorus and their followers[10] – claimed that place is void, others that it is extension.[11] Having claimed just this much, that place is void or some kind of extension, they left it at that.

After these remarks, Aristotle embarks on his actual investigation of place. First he shows that place exists, immediately afterwards arguing for the contrary, that it does not exist, so that we may know, on the basis of the arguments on both sides, which arguments concerning place are sound and which ones are not, and how we should and should not conceive of place.

He begins by showing that place exists, on the strength of five arguments, the first of which is based on the mutual replacement of bodies. For, he argues, where there was water, there, as the water receded, there came to be air. And when air in its turn has moved out, then another body comes to be in the same place, as happens also in the case of vessels. This being so, Aristotle will form the following syllogism about place: (*a*) place can receive different bodies in turn, while remaining one and the same; (*b*) that which, while remaining one and the same, can receive different things in turn, both is something itself and is different from what it receives; therefore (*c*) place both is something itself and is different from what it receives.[12] For that which is in no respect or way existent could not receive anything.

On the strength of this argument he shows that place is something, but on the basis of the second argument he shows not only that 499,1
place exists, but also that it has certain natural powers and differentiations.[13] The argument takes its start in the tendencies of natural bodies. For each natural body, he says, when not impeded, moves to some definite place: light bodies upward, heavy ones downward. Whenever light ones move downward or heavy ones upward, they move contrary to nature and by force. Now if some things by nature move upward, and other things by nature move downward, it follows that 'up' and 'down' exist by nature, and these are differentiations of place – up and down, right and left, and before and behind – so that place not only is something, but also comprises natural differentiations. What is more, each of these places also has a cognate power.[14] For if everything seeks its natural place, it is clear that for everything its natural place must be an object of striving and what it seeks. But an object of striving is such through its having a certain natural power. For that is what is being sought. But if someone says that light bodies do not seek the place above, nor heavy ones the place below, but instead seek to obtain the totality to which they belong,[15] he should know that, if this were the case, heavy things should not move exclusively at right angles, but also in any other way whatsoever. For it would be possible, with plenty of earth around, for heavy things not to get separated from the totality to which they belong even if they did not move downward exactly perpendicularly. But as things are, since their natural tendency is towards the centre, this is why they move at right angles exclusively. When a stone falls down into a deep pit, even if it is launched sideways towards the totality to

which it belongs – for we take it to be surrounded by earth on all sides – it will still continue its movement downwards perpendicularly only. And if some part were to break off from the wall of the pit, this again would move downwards in the same way. Thus all things heavy have a tendency to move towards the centre, not just towards the totality to which they belong; and even in so far as they do in fact move towards the totality to which they belong, they move towards it because this totality has obtained its natural place. And we shall affirm the same things about things that are light. For these too have one single natural place, the concave surface of the sphere of the moon, towards which they too move at straight, or equal, angles. So from this second argument too we obtain both that place is something and that it has natural powers and differentiations. For its directions are the six that have been mentioned. Moreover, Aristotle claims, these directions surely do not exist in a way that is merely relative and as relates to us. For we do also use these distinctions relatively to our position – hence the same person may come to be to the right and to the left, and the same thing may come to be in front of us and behind us, if the position of some thing is changed; or above and below, as with a roof. For what lies on a roof has the roof below it; but once it slides down and gets under the roof, what was formerly below is now above.[16] But even if above and below and the other directions also exist relatively to us, they are nevertheless also there in the universe by nature, distinguished in accordance with the motions of the natural bodies. For we call 'below' the place to which heavy bodies move by nature, and 'above' the place to which the light ones move. In *On the Heaven* Aristotle shows that right and left too exist in the cosmos by nature, describing the right as where the starting points of motions are – e.g. the points at which the heavenly bodies rise – since in us too these are on the right side, and describing left as what lies opposite to this.[17] Analogously, 'in front' will be everything in the north, and 'behind' the things in the south. In case someone should not accept this, he will at least need to grant that 'above' and 'below' exist by nature, and these are varieties of place, so that it still follows both that place exists and that it comprises natural differentiations.

The third argument takes its starting point in a consideration of mathematical magnitudes.[18] That place exists by nature, he says, is something that is clearly shown by mathematical magnitudes.[19] For since every magnitude consists of parts that have a particular position, and mathematical magnitudes are like this as well, these magnitudes as well have a certain position and are somewhere. For we say that of this triangle this here is the base, this here the top, and this is the right side, and this the left side, but their places are relative and only exist in our thought. Hence if I turn the triangle around, I thereby create another top, and another base, and I switch the position of the sides. And this is only reasonable, because these things do not exist in their own right, but have their being in our thought. And just as they have their being in our thought and not in nature, they also have their places there. Now if these things that exist in thought also have their places in thought, we may conclude that if something similar exists by nature, its parts will have their position and their places by nature.

These arguments are based on actual facts. The remaining two, however, are based on received opinions and on the testimony of earlier thinkers. The fourth one comes from those who introduce the void – if, he says, there is such a thing as void, and these people themselves claim that void is nothing but place bereft of body, it is clear that place is among the things that exist. The fifth one is that Hesiod not only says that place exists, but also that it has come to subsist, by nature prior to all other things: 'First of all', he says, 'Chaos came into being, and then broad-bosomed Earth.'[20] For it looks as if he as well, Aristotle claims, is among the people who think that there is no existing thing that is not in place. This at least is why he says that place came to subsist before all other things, calling it 'chaos' because it is the room (*khôra*) and that which provides room (*khôrêtikon*) for all other things. In that case it seems to follow, Aristotle claims, not only that there is such a thing as the nature of place, but also that it is an amazing thing and prior to everything else. For if it disappears it removes everything else with it, but it does not itself disappear along with anything else. For if there were no place, nothing else could exist, since all things that exist are in a place; and if this is not true,[21] then at least bodies could not exist in the absence of a place for them to be in. But nothing prevents place from existing in the absence of bodies. And even if one did not want to accept that place is prior in time, it is in any event necessary that it is prior at least by nature.[22]

> **208a27** The student of nature must have knowledge about place too, just as he must about the infinite, whether it exists or not.

In this very same way he also made his enquiry into the infinite: he argued on both sides and then showed in what sense it does not exist and in what sense it does.

> **208a27** and in what sense it exists and what it is.

In what sense it exists: for example whether it exists in its own right, as was thought by those who took place to be the void or an extension, or whether it has its existence in something else, as Aristotle believes (for in his view place is not a separate subsistent thing existing in its own right). And **what it is**: whether it is void, or extension, or the limit of the surrounding body.

208a29 For they all suppose that the things that exist {are somewhere}.

{He says} that the theory of place is thus necessary in so far as it is an opinion shared by all people that if something exists, it is by all means in a place, so that he who wants to get to know the things that exist must know place too, without which these things cannot exist. And they establish this – i.e. that everything that exists is in a place – by incorrectly applying a conversion with a negation. For assuming
502,1 that what does not exist is nowhere – a true assumption – they convert the argument incorrectly on the basis of {the negation of} the antecedent: what is, therefore, is somewhere.

208b1 That place exists, seems to be clear from replacement.

Here starts the first argument establishing that place exists. **For where there is water, there, when the water has gone out as if from a vessel, air will be in its turn**. This is the minor premiss of the syllogism, viz. that place, while remaining one and the same, receives ever-different bodies. **This, then, seems to be different from all the things that come to be in it and that move about**. The major premiss: that which while existing as one and the same receives ever-different things, both exists and is different from the things it receives. **So it is clear that place and space must be something other than either**. This is the conclusion of the syllogism.

208b8 Again, the locomotion of the simple natural bodies.

The second argument is that on the basis of natural tendencies. He fittingly concentrates his argument on the simple bodies. For the composite bodies also have their natural motions according to the respective predominance of the simple ones.

208b12 These are parts and kinds of place.

He uses the terms indifferently, because it is indifferent at this point whether these directions are parts or kinds of place as a whole. For this does not contribute to the question at stake. After all, we are at this point investigating whether place exists at all or not. Those who claim that place is extension or void are likely to describe them as 'parts'; those who claim that it is the boundary of what surrounds will call them 'kinds'.

208b19 For 'above' is not anything you like, but where fire moves to.

That, in the natural world, 'above' and 'below' are not just relative, but exist by nature and are fixed, is also clear from the following. 503,1
Suppose there is a house with a roof in the middle, then if this is by its relative position both 'above' and 'below', fire, if let loose from below, or a stone, if let loose from above, will not stop in the middle, but the one will move upward and the other downward, when they are not impeded but can move freely through space.

208b21 So that they differ not only by relative position, but in power too.

For if each of these places is receptive of a maximally different kind of body, it is clear that they differ from each other not only in their relative position, but also in their power. For either one of them is an object of striving, the one for light bodies, the other for heavy ones.

208b22 Mathematical objects, too, make this clear.

The third argument: if these objects, which are not in a place since they do not naturally exist in their own right, nevertheless when they are made existent in thought immediately acquire a position and a place which is in conformity with their way of existence – I mean a place which also exists in thought – it is much <more>[23] the case that physical bodies that are indeed existent will have a place that is physical as well, and that exists not relative to someone or something, but by nature. For it is necessary that according as the emplaced body is disposed, so place must be disposed as well. If what is emplaced exists in thought, then its place will exist in thought. If what is emplaced exists relatively to something, so will its place. Hence also for things that are naturally somewhere, their place too will exist naturally.

208b25 Again, those who claim that there is void are talking about place.

This is the fourth argument, which is based on opinion, not compelling, just as is the fifth.

208b29 Hesiod, too, might seem to be speaking correctly in making Chaos first.

The fifth is based on the testimony of Hesiod.

208b33 If it is something like that, the power of place will be 504,1
remarkable and prior to all other things.

If, he says, place is of such a kind as Hesiod and others claim it is, its nature is surely remarkable, and it is not just among things that exist, but prior to everything else. For if other things do not exist without it, whereas it itself exists prior to everything else, it is clear that it is prior both in nature and in time. For if it disappears it removes everything else with it, but it doesn't itself disappear along with anything else.

> **209a2** For all that, it is a problem: if it is, what it is, whether some kind of bulk of body or some other kind of thing.[24]

Having shown by means of five arguments that place exists, he goes on to show what it is – for that is the second problem. However, through his examination of what it is, he brings around the argument to showing that place does not exist at all, in order to make clear, as I have said, that the account of place is difficult and that the arguments on both sides of the question are such as to exercise the mind and to provide an exact understanding of the subject matter, not only in so far as the problem itself gets clearly delineated, but also in so far as the accounts that try to destroy it {i.e. place}[25] are being refuted.

He tries to shake our presupposition about the existence of place through several arguments. It is not to be demanded that the accounts are cogent on both sides, for that is impossible. However that they are plausible, not that they are cogent, is what one should at any rate demand. So one should not be surprised, if some of the arguments indeed do achieve plausibility. However, only then can the arguments have cogency on both sides, when they take what is not true of the object to be true, and construct their syllogisms on that basis.[26]

As I have said, he starts his account with an investigation into what place is. Since we acquire understanding of what a particular thing is by means of definitions, and definitions consist of genera and differentiae, we must first, or so he claims, find the genus of place. Now it is necessary that place should be either a body or incorporeal. First, then, he shows that it is not incorporeal, by showing that place
505,1 must be a body. If the body, in so far as it is body, is in a place, it is clear that it must be in place with every one of its dimensions. For suppose that its dimension of length is in a place, but some other dimension is not, then it is not in a place in so far as it is body, but in so far as it possesses a certain length or breadth. But if the body is in a place in so far as it is a body, it will obviously be in a place with its three dimensions. Hence the place that is capable of receiving this body must be three-dimensionally extended as well. For if it is not three-dimensionally extended, but for instance extended in just one or two dimensions, it does not encompass the three dimensions of the

body. Hence the place of the body must be three-dimensionally extended, and everything that is three-dimensionally extended is a body – for this is the definition of a body: what is three-dimensionally extended – and so it turns out that place is a body.[27] The syllogism is of the first figure: place is three-dimensionally extended; everything that is three-dimensionally extended is a body; therefore place is a body.[28]

Having thus shown that place must necessarily be a body, and having thus ruled out that it is an incorporeal, he goes on to show, in turn, that it is not possible that place be a body either. For suppose place is a body, then there will be two bodies in the same place. But it is impossible for two bodies to be in the same place; therefore it is impossible for place to be body – all this according to the second mode of the hypothetical syllogism, which by ruling out the consequent, rules out the antecedent.[29]

But why would we assume that the first inference is true – i.e. that if place is a body, there will be two bodies in the same place? This you will establish through a categorical syllogism, as follows: every body is in a place; every place is a body; hence every body will be in a body; but if a body is in a body, two bodies will be together in the same place. That it is impossible for two bodies to be in the same place is what you can conclude from the following argument.[30] If it were possible for a body to completely pervade another body, so as for the two of them to be together in the same place, then it would be possible for the largest thing to be contained in the smallest, and, as Aristotle himself puts it, the sea could be contained in a ladle and the heavens in a millet-seed.[31] For it is possible to cut up the largest possible substance into very small equal portions, for example the sea into portions of a ladleful. Now if the same place can contain two equal portions, why not also three or four or an infinite number? For where would one decide to stop? If this is the case, then the sea will be contained in a ladle, and the heavens in a millet-seed. After all, every limited quantity is exhausted if limited parts of equal size of whatever kind are being subtracted. And if this is impossible, then it will also be impossible for two bodies to be in the same place. Furthermore, if a body can pervade another body, then a cupful of water added to two cupfuls of wine should not make three cupfuls in all, but two, so that the same bulk of the two cupfuls of wine would be preserved. For if it is not just the qualities that pervade each other, but the bodies themselves do this as well – for example the water goes through the whole substance of the wine – then the combination of the two should not need to take up more space. Alternatively, if the cupful of water completely interpenetrating the two cupfuls of wine does in fact make the whole increase in size, the combination should measure four cups in bulk, given that the one cupful has spilt out into the bulk of two cupfuls; or, conversely, when two cups of wine are added to the cupful of water, the combination should measure just one cup, since the two cupfuls have pervaded the single one.

If, then, it is impossible for two bodies to occupy one and the same place, since it is impossible for one body to pervade another one, then it is impossible for place to be a body. But neither is it possible for it to be incorporeal, as has been shown. For it will either be completely dimensionless, and thus fail to[32] receive what has dimensions, or have only one or two dimensions, and thus fail to receive the whole body in its own totality;[33] yet the body needs to be emplaced in so far as it is a body. Now if it is not possible for place to be either a body or incorporeal, then it appears that place does not exist at all. For if it did exist, it would by all means be either a body or incorporeal.

Next, he says, it will be possible to prove, by the very same arguments by which we have proved that place exists, that it does not exist. Its existence was proved first and foremost through the first argument, which went like this: if place, while remaining one and the same, receives ever different bodies, and if that which, while remaining one and the same thing, receives different bodies both exists and is different from the bodies it receives, then we must conclude that place both exists and is different from the bodies it receives. Now by this very same argument it is possible to show that place does not exist. For it is clear that, just as where formerly there was the water, there wine has come to be afterwards, or air, or something else – just so where formerly there was the surface of the
507,1 water, there the surface of the wine has come to be after the water had moved out; and the surface of the air, after the wine had moved out. Hence there also exists a place of a surface. For being one and the same, it receives ever different surfaces. In the same way we will show that there also exists a place of a line. For if there is a place of a surface, it is clear that there will also be a place of its limit, i.e. a line. For where formerly there was the limit of the surface of the water, there later on the limit of the surface of the air has come to be, and then that of the wine, and so on. Hence there exists a place of a line. But if of a line, then also of its limit, namely the point, according to the same argument. Yet, since the place is one thing and that which is emplaced is another, the point will evidently be different from the place of the point, which of course is a point itself as well.

However, one would find it hard even to conceive what the difference should be between a point and its place-point.[34] It is possible to specify some difference between a body and the place of a body, even in the case where place is a body, because the place-body is unqualified, whereas the emplaced body possesses qualities. But between a point and a place-point it is not even possible to conceive of a difference. It would have been possible to make the same remark also about surface and line, but Aristotle himself confines his argument to the point, as the clearer case.[35]

Alternatively:[36] if there is a place of a point, then, since the main differentiations of place are these two: up and down, and since the other four differentiations are due to these two, and since it is impossible even to conceive of any additional differentiation of place, it is clear that also the place of the point would be differentiated according to one of these differentiations and that it will either by nature be up or by nature be down, so that also the point emplaced in it will by nature be either heavy or light (for what is by nature in the place high-up, is light, and heavy is what is by nature down below). But it is impossible for a point to be heavy or light. It follows that it is impossible that there should be a place of a point.

But if there is no place of a point, then neither of a line. For what encompasses anything will evidently also encompass the thing's limit, since otherwise it would not even encompass the whole thing itself as a whole either. Hence, if there is no place of a point, neither is there of a line; if there is no place of a line, neither is there of a surface; if there is no place of a surface, neither is there of a body; and if there is no place of a body, then neither is there a place of any other physical object,[37] let alone of anything else. So it seems that there is no such thing as place at all. So this is the second argument by which place has been shown not to exist.[38]

Next, he says, if place is among the things that exist, then, since all existing things are either intelligible or sensible, and among the sensible things some are elements and the others composed of elements, while among the intelligible things again some are elements, others composed of elements, place too must necessarily fall under one of these headings, and must be either intelligible or sensible, and in either case be either an element or composed of elements. But it is impossible that place is any one of these things. For of sensible things, i.e. the physical bodies, the elements that are actual and exist are themselves bodies as well – say, the four familiar ones – and it has been shown that it is impossible for place to be a body. Hence, place is not one of the sensible things, neither as an element nor as something composed of elements. But neither can place be among the intelligibles. For in the case of the intelligibles the elements are intelligible as well (the elements of syllogisms are premisses, and of these the elements are syllables,[39] and both of these are intelligible; and the same holds for all objects of intellection), and these are completely without extension and magnitude (for they are neither lines, nor surfaces, nor three-dimensional). But place, if it exists at all, is an extension, since it is also receptive of extensions. Hence place is not one of the intelligibles either. Now if place turns out to be neither intelligible, nor sensible, and all existing things are either intelligible or sensible, then it would seem that place is not among the things that exist.

Suppose someone says: what is it that prevents place from being an element of sensible things without being a body itself, since both matter and form are also elements of sensible things while being incorporeal? I reply that, in the first place, it has been stated that it is impossible for place to be extensionless while being the space for extensions.[40] Subsequently, through the next argument, the fourth, it will be shown at a more general level that not only is it impossible for place to exist as an element, but that in general it is not possible for it to exist as a principle in any of the senses of this word. For since he has shown that it cannot be either an element or composed of elements, and since all the things that exist are not just elements and things composed of elements,[41] but there are also other things besides these – for the productive and final causes are neither elements nor in any way composed of elements, but they exist as principles, so that the most general division of things that exist is one into principles and things that are derived from principles – and since he has shown that it is impossible for place to be derived from principles (for then it will be either intelligible or a body, and it can be none of these), he now accordingly wants to show that place is not in any way a principle and cause of things that exist either.[42]

For the causes of all existing things, he says, are four in number – the matter, the form, the productive cause, the final cause – and place cannot be any one of these. For nothing comes to be from it, but all things come to be from matter; and place contains, but matter is being contained. Hence place cannot exist as matter. But neither is place form. For the form is what completely determines the essence of each thing, and each thing that exists is characterised according to it, and that which loses it has perished immediately (for example the form of fire is the hot and the dry, and if fire loses it, it no longer continues to be fire, but perishes). But a thing that leaves its proper and natural place does not exist any less and preserves its proper form. For fire, both when it is down below and when it is high up, is none the less fire, and when it arrives down below, it has left the place that belongs to it behind, but not its fiery form. So place cannot exist as form either. Nor, however, as a productive cause. For we do not see anything being created by place. Nor as a final cause (which is actually surprising; after all place appears to be like an end or an object of striving).[43] For that which is striving for something wants to become that which it is striving for – for example what is striving for the good wants to become good, and what is striving for health wants to become healthy. But none of the things that are in place actually becomes place.[44] So place is not a final cause either.[45] Therefore, place cannot be any of the causes. But it has been shown that it cannot be any of the things that are produced by a cause either. Therefore place appears not to exist at all.

With respect to this {last argument} one might raise the following difficulty. If what is striving for the good wants to be *in* a good state,

and if what is striving for health wants to be *in* health, and if similarly what is striving for a particular place wants to be *in* that place, why is it not the case that, just as the good and health are the end of the things that strive for them, so also place is the end of the things that want to be in it?[46] For certainly everything is at rest and stops striving as soon as it has reached that which was the object of its striving, just as things in locomotion also stop moving once they have reached their natural place, which is indeed what they are striving for. So how come, then, that their place is not a final cause? I reply, first, as has already been said, that things striving for something, when they have obtained it, in so far as is possible for them become that which is the object of their striving, and are named after it, and in some cases even share its name (certainly, he who strives for the good and obtains it, is good and is called good, and he who is striving for health healthy, and so on), but that which is striving for a place neither becomes place, nor is called after it (for even if he who strives after virtue is as a rule not named after virtue, he at least to some extent becomes precisely that which virtue is). Also otherwise: final causes are seen to be present in the things of which they are the ends, but place is different from all the things that 510,1
are in it, having no share in the emplaced object.

Next, he says, also **Zeno's puzzle requires an explanation** and a defence.[47] For if all things that exist are in a place, as some people have thought, and if place as well is among the things that exist, then place also will be in a place, and the latter again in another one, and so on *ad infinitum*. This is the fifth argument.

Sixthly and finally, he comes up with the argument from things that grow. If, he says, there can neither be a body without a place (given that people say that all things are in a place), nor place empty of body (for a little further on it is shown that void cannot exist), what shall we say about things that grow? When a body that grows gets bigger, whence will its place come? For everything is already filled up, because there is no void. It is necessary, therefore, that also its place grows bigger, and that it grows along with the body. But then it is hard to see how and whence this place will grow. And if the thing that grows covers a larger place, its place also will cover a larger place, and once more the place will be in a place. Furthermore, the place it has occupied was either empty before or filled with body. But it is neither possible that it was empty, nor that there were two places and two bodies together in the same place. The remaining possibility is that what has been added to the body that has grown bigger is not in place, which is itself impossible as well. So if bodies grow, and if, given that place exists, there can neither be void nor a body that is not in a place, and if either of these must necessarily be the case when bodies grow, it would appear that place does not exist at all.

209a2 For all that, it is a problem: if it is, what it is.

Even if we have proved that place exists, he says, still the account of what it is contains many puzzles, so that from that point onward our account comes round to the position that it does not exist at all.

209a4 It has three dimensions: length, width, height, by which every body is bounded.

By **it has three dimensions** he indicated the minor premiss, and
511,1 through **by which every body is bounded** the major. The minor was: 'place is three-dimensionally extended'; the major: 'what is three-dimensionally extended is a body'; he has skipped the conclusion as being obvious (it was: 'therefore, place is a body'). And the absurd conclusion that follows from this argument, being the conclusion of the second, hypothetical, syllogism he presents thus: **it is impossible for place to be a body**. The hypothetical syllogism looked like this: if place is a body, two bodies will be in the same place; but it is impossible for two bodies to be in the same place; therefore it is also impossible for place to be a body. Of this syllogism he has omitted the composite premiss 'if place is a body, two bodies will be in the same place', and presented just the second premiss and the conclusion, giving first the conclusion by saying **it is impossible for place to be body** and then the second premiss by saying **for then two bodies would be in the same place**, which is equivalent to 'it is impossible for two bodies to be in the same place'.

209a7 Again, if there is a place and a space of a body.

The second argument.

209a9 For the same argument will apply.

The same argument, he says, which he adduces concerning the mutual replacement of bodies will also apply in order to show that, if there is a place of a body, there will also be one of a surface and a line and a point.

209a12 So that if not even the place of that is different {from the point itself}, then neither will {the place} of any one of the others be.

If the place of a point, he says, is no different {from the point itself}, evidently neither will the place of a line, or a surface or a body be. How come the place of a point is not different from the point? For in what respect will the place of a point actually differ from the point

itself? For it is not larger. After all, in so far as a part of it exceeds it, in so far will it not contain any of it and in that respect it is not the place. When it is equal, then it will be a point in that very respect, so that the whole thing will be just one point. For the points coincide. For even if ten thousands of them are made to coincide, they do not 512,1
make the whole thing larger. And if the point is {thus} one,[48] the place-point and the emplaced point will not be different. The same applies to a line and a surface. And if there is no place of any of these, there will not be a place of a body either. For what is in place must be in place in its entirety, just as the wine in a vessel.[49]

209a13 For what could we ever suppose place to be? For place can be neither an element nor made up of elements.

This is the third argument.

209a15 Having a nature like this.

This stands for: being receptive of bodies, remaining one and the same without change and contributing nothing to the nature of the {emplaced} bodies. This he says because Plato seems to call place matter, a thesis which he will disprove later on.[50] So if, he says, we stick to the true conception of place, which we naturally have, without using the term in an unnatural way, it is difficult to formulate the essence of place.[51]

209a18 From intelligible elements no magnitude comes to be.

Certainly matter and form are non-sensible elements (for what is sensible is the composite), and yet from their combination as intelligible things magnitude does come to be. But these have never come to subsist in actuality on their own, whereas we are talking here about things that actually subsist, and place too subsists. Anyway, he does not affirm this about every intelligible element without further qualification, but about those elements that are elements of intelligible things, while of course being themselves intelligible as well. From these then, he says, when we put them together, no magnitude comes to be – for example: from terms or from premisses; for what comes to be from these, the syllogism, is intelligible too.

209a18 And further, what could one suppose place to be the cause of for existing things?

After he has said, in the constructive overview of arguments about place, that place has a considerable power, in the present argument he wants to establish the opposite idea, namely that it has no effect on existing things. For it cannot be reduced to any of the four causes.

513,1 **209a22** nor as a form and definition of things.

Form and **definition** are juxtaposed. For the form of each thing is no different from the definition expressing the 'what it was for this thing to be'.

209a23 And also if place itself is one of the things that are, where will it be?

Another argument. If, he says, all the things that exist, are in a place, and also place is among the things that exist, then place also will be in a place, and so on *ad infinitum*.

209a23-4 For Zeno's puzzle requires an explanation.

– that is: a solution and a defence. Perhaps Zeno came up with this puzzle and removed the manifold variety of places from reality by removing place in general, with a view to showing that what is is one. For the variety of places presents us very clearly with the plurality of things. This we say by way of conjecture. For we cannot say with any certainty what Zeno wanted to achieve by setting up this puzzle. It may also be that he wanted to introduce the infinite. For if place is in a place, then that will evidently be in another, and so on *ad infinitum*. Hence, the infinite exists. Or, if one wants to give a charitable reading, perhaps it was in order to demolish the argument that all things that exist are in a place that Zeno said this, viz. that if all the things that exist are in place, and place is among the things that exist, place also will be in a place, and that in another one, and so on *ad infinitum*. So that it would follow that if the latter conclusion is strange and impossible, then it turns out that not all things that exist are in a place.

209a26 Again, just as every body is in a place, so also there is a body in every place.

The final argument. He assumes these two self-evident principles: one, that every body is in a place (for that is generally agreed by all people, given the fact that they even said that all things that exist are in a place); the other he takes to follow from this, viz. that there is a body in every place. After all, place is being spoken of relatively to what is in place, and he will also show later on that there is no void at all. These things being so, he says, what shall we say about things
514,1 that grow? For either the body that is added will not be in place,

which is just what is impossible according to those who introduce {the notion of} place; or it finds itself in some kind of pre-existing void, and this too will be shown to be impossible; or place has grown and from being small has become larger, and this too is ridiculous. For what kind of growth could anyone conceive of in the case of place?

From all this then, he says, it may be concluded that not only have we not found what place is, which is what we were supposed to show here, but we have come round to showing that place does not even exist at all.

> **209a31** Since things are spoken of on the one hand in respect of themselves and on the other in respect of something else; and place is on the one hand common, that in which all bodies are, and on the other particular, that in which a body is primarily ...[52]

After having told us in what precedes that on the basis of the properties that apply to place one cannot formulate a single account of it, but that on the basis of some properties place might be thought to be the form of each thing and on the basis of other properties the matter, he now wants to show precisely this: on the basis of which properties it might be thought to be matter and on the basis of which, form. Next, he shows, by means of several arguments, that it can be neither matter nor form.

In order to show how it might be thought to be form and how matter, he also does now what he is wont to do, i.e. he distinguishes the different meanings of 'place', and specifies what is the place in the proper sense about which we are talking now. For on other occasions too he is used to setting out the different meanings of words that are homonymous, and to specify the subject matter of the investigation at hand.[53] This is what he also does now, as a necessary prerequisite for the person who gives an account of place and, besides, in order to show that on the basis of the properties of place it sometimes will appear to be form, sometimes matter; for place understood in the sense of the more common kind does not appear to be thought of as form.[54]

To have us find out in how many senses the term 'place' is used, he uses a distinction of the following kind, viz. that where some things are something in the primary sense, others in respect of another, a thing's place too can on the one hand be the place of the relevant thing in the primary and most proper sense, and on the other hand the place of this thing in respect of something else. For example, man is said to be capable of sense-perception, but also receptive of understanding and knowledge; but he is capable of sense-perception in respect of something else (for he is so *qua* animal, not *qua* man), but receptive of understanding and knowledge *qua* man. And in the case of numerous other things one can make the same distinction. Place too, then, will in the one case be primary and in its own right, in another case it will be in respect of something else. Thus, I can be said to be in the heavens, he claims, not in the primary sense, but in respect of something else, in virtue of the fact that the heavens contain everything. Surely I am not in the heavens, but I am 515,1 in the air, and in virtue of the fact that the air is in the heavens I too am said to be in the heavens. It is in respect of something else, then, that I am in the heavens, not in the primary sense. Similarly I am said to be in the air, not in the primary sense, but in virtue of the fact that I am in this or that particular part of the air. And I am said to be on the earth, not in the primary sense, but in virtue of the fact that I am standing on just that part of it which, together with this part of the air here, immediately surrounds only me. And this is what in the primary and most proper sense is the place of each thing: the place that surrounds just the one individual and separates this off from the other things. So, to put it in a definition, place in the proper sense is the place that immediately surrounds each individual body. And if that which surrounds each thing is its place, then, he says, **place will be a kind of limit** (209b2). But if place is that which primarily limits and defines each thing, place would seem to be its form. For this is the distinctive property of a thing's form: to define and limit the individual thing and to set it off from other things.

On the basis of these characteristics, he says, the place of an individual thing might be thought to be its form, but on the basis of other characteristics of place again it could be held to be matter. For if place is an extension providing room for, and receptive of, magnitudes, it would seem to be matter. For matter is such a thing, being in itself an indefinite extension, and it is defined by the surfaces and outlines and other properties of the magnitude. For if, he says, from a sphere we take away its properties, outline, and limits, nothing is left of it other than a certain indefinite and limitless extension, and this is the distended matter.[55] Since place too seems to be something like this – for it is a kind of extension which in itself is without form or quality, but is receptive of things that possess form and quality – in this respect it could also seem to be matter. And not in this respect only, but also on the basis of the fact that while remaining one and the same it is receptive of ever different magnitudes. For this is very much a proper characteristic of matter.

It is for these reasons, he says, that Plato too in the *Timaeus* claims that the participant and space are one and the same thing, and the participant for him means: matter.[56] Now if matter is the participant and the participant is space, and space is nothing other than place, then Plato was saying that place is matter. If he gave a different name to matter in the *Timaeus*, calling it the participant, and if he calls it 'the great-and-small' in his unwritten teachings, this makes no difference to us, the relevant thing being that he identifies the participant with place and space.[57]

About place all the others, he says, merely said that it exists, but 516,1
only Plato attempted to say what it is, telling us that matter is space and place. Now Aristotle, as is his wont, is here examining the outward appearance {of the argument} and in that sense reasonably takes Plato to task for saying that matter is place.[58] However, it is very clear that it is not the kind of place we are talking about here now – i.e. the place that can receive compound bodies – that Plato called matter. It is rather the place of the physical forms that he called matter, by analogy, because just as every body is in a place, so every physical form is in matter.[59] This is similar, indeed, to the way in which Aristotle himself too in his *On the Soul* calls the soul the place of the forms: 'and those people put it very well who say that the soul is the place of the forms', and he says that the intellect is the place of the intelligible forms.[60] It is in this way that also Plato says that matter is the place of the physical forms. For if, just as the psychic forms stand to the soul, so also the physical forms stand to matter, and if soul is the place of psychic forms, then also matter will naturally be the place of the physical forms, in the sense that it can receive them and that they cannot be sustained without it. For the separate and transcendent form is analogous to the father, and matter to the mother; the demiurgic patterns that are implanted in matter from the separate forms are analogous to the seed, and the physical forms that blossom up in matter from the demiurgic patterns are analogous to the foetus.[61] This is why the *Timaeus* called matter the 'mother', the 'nurse' and the 'receptacle', and why Plato also calls matter the participant, because it participates in all the forms and receives them, and only plays the participating role, while in no way being participated in. In what sense he called matter the 'great-and-small' has often been explained: because it is receptive of contraries, and the first and most generic opposition between contraries is that between great and small, since matter first acquires volume by being quantified, and the first opposition in the category of quantity is that between great and small.[62]

Having thus shown, from a survey of the properties that belong to place, on the basis of which properties it might seem to be matter and on the basis of which form, he now goes on to show, through various arguments, that it is neither possible for place to be matter, nor to be form. If we count the arguments, they are eight in all.

Through the first argument he shows that place can be neither
matter nor form, discussing them together. The argument goes like 517,1
this. Both matter and form, he says, cannot be separated from the thing of which they are the matter and form. Place is separate from the things of which it is said to be the place. Therefore place is neither form, nor matter. That the things are separate from the places they are in, is evident. For in the place where formerly there was water, he says, there air has come to be, when the water has moved out, just

as in a vessel in which there formerly was wine, there, when the wine has moved out, water comes in, or air. In fact the vessel, he says, is a portable place, just as place is like an immobile vessel. Just as the vessel is nothing of what is in it – neither its matter nor its form, nor anything else (because matter, form and other such things are inseparable from the thing itself, but the piece of pottery is separable) – just so, I say in general, it is with place.

The second argument is presented in two parts. By the same argument {as the one just recorded} he shows that place is not form, given that the form is inseparable and place separable. But that place is not matter either, he shows as follows. Place contains the things that are in place. Matter is contained and does not itself contain. Therefore place is not matter.

The third argument goes like this. Place is external to the thing of which it is the place. Matter and form are not external to the thing to which they belong. Therefore place is neither matter nor form.

The fourth argument proceeds through a hypothetical syllogism, as follows. If matter and form were place, nothing would move to its proper place (for nothing moves towards that in which it already is;[63] and each thing is in its own matter and form). Yet all things do move to their proper places. Therefore matter and form are not place.

The same argument can be transformed into a categorical syllogism, as a fifth argument, as follows. The things that exhibit natural motions move towards their proper places. But none of the things that move moves towards its matter or form. Therefore place is neither matter nor form. The syllogism is of the third figure.[64] That nothing moves towards its matter, or its form, is evident. For a thing moves towards that which it has not yet reached before the movement starts; but it has all its being-that-very-thing in its matter and its form.

The sixth argument. In place there is above and below, towards which things moving do indeed move. But in matter and form there is no above and below. Therefore place is neither matter nor form.

518,1 The seventh argument goes like this. If, he says, form or matter are place, and matter and form are in the thing, i.e. in their compound, then, if the thing moves from place to place, evidently they too move along with the thing and arrive in the same place as it. This means that place too moves along with the things from place to place. Now first of all it is absurd that places move. Secondly, it will thus turn out that place is in a place. For if the compound comes to be in a place when it moves from here to there, and if the compound is nothing other than matter and form, then it is clear that matter and form come to be in a place as well. But matter and form are place. Therefore place comes to be in a place. You can reduce the argument to a syllogism of the first figure as follows: place is either matter or form; matter and form are in the compound (for the compound is nothing other than matter and form); the compound leaving one place comes to be in another place; therefore place too leaving one place comes to be in another place. And if it is absurd that a place comes to be in a place, then place is neither matter nor form; for it was on this hypothesis that the absurdity followed.

The eighth argument. If form is place, he says, then when water perishes and changes into air, the water's place necessarily perishes too, given that form is place. But who could envisage the perishing of place? For just as there is no such thing as a coming-to-be of place, so there can be no such thing as a perishing of place. Apart from this, neither does sense-perception suggest that the water's place is gone when the water changes into air, for in this place another body has come to be. And on the supposition that matter is place: since matter remains the same when the water changes into air, why does water when it changes into air change place even though the matter is still the same? All these arguments make it clear that place cannot be either matter or form.

Chapter 2

> **209a31** Since things are spoken of on the one hand in respect of themselves and on the other in respect of something else.

In respect of themselves is opposed to 'incidentally', and 'primarily' to **in respect of something else**. Here, then, he uses 'in respect of itself' as equivalent to 'primarily'.[65] Since he wants to assimilate place to form, on the grounds that, just like form, place too circumscribes and defines the individual thing, and since place is in one sense understood to be common, whereas in the other sense it is proper to the individual thing, and since the resemblance with form does not concern common place, but proper place – for these reasons he says that since some things are said to be what they are in a primary sense and others in respect of something else, evidently place too will be on the one hand, in the primary and most proper sense, {the place} which surrounds the individual thing individually, and on the other hand, in respect of something else, the common {place}. We are said to be in the cosmos because we are in this 519,1
particular part of the cosmos, and we are in that part because we are in this city, and we are in that city because we are in this house, and in that house because I am in the part of the house that immediately surrounds just me, which is in the proper and primary sense my place. For the house, the city, and the cosmos are not in the primary sense my place, but in respect of something else: the house is my place owing to the part of the house that surrounds me, the city owing to the house, and so on.

209b1 If then place is that which primarily surrounds each individual body, it will be some kind of limit.

{Place,} that is: what primarily and immediately surrounds the individual thing, not the common place. If place is what immediately surrounds the individual thing, place is evidently a limit. For it limits what is in it. And form is also like that.

209b2 So place would seem to be the form by which the magnitude and the matter of the magnitude are defined; for this is the limit of each thing.

By **the magnitude** he means the compound of matter and form. He says, then, that the compound too is defined by the form; for the fact that it is a definite thing is due to its form. And the matter of the compound, which is indefinite in itself, is being defined by the form. For the form, he says, is the limit both of the matter and of the compound. **Form** he now appears to use in the sense of the limiting contours. For also the extension {of the compound}, being indefinite and limitless in itself, is limited and defined by the surfaces.

209b6 In so far as place seems to be the extension of the magnitude, {it seems to be} matter; for this is different from the magnitude.

That is: in so far as place is an extension that can receive the magnitude, in that respect it seems to be matter. For in the magnitude this indefinite extension is the matter of the magnitude. Prime matter, by receiving ever different forms on the one hand, shows itself to be different from these forms, but it is never without form. Just so,
520,1 I claim, since the magnitude sometimes becomes larger, sometimes smaller, it has something different underlying it, namely magnitude *simpliciter*, i.e. an extension that is indefinite on its own account, and which Plato calls 'in a process of discordant and disorderly motion'.[66] But this can never be found without a particular boundary and limit, even if it has on its own account no boundary and no limit. Now just as this indefinite extension, being the matter of the magnitudes, receives ever different magnitudes,[67] so place too is an extension receptive of ever different magnitudes, and in that sense it might seem to be matter. And just as the indefinite extension is different from the magnitude, e.g. from the two-cubits-long or three-cubits-long, so the place is different from the magnitudes that come into it.

209b7 This is that which is surrounded and defined by the form, that is to say by plane and limit.

For the indefinite extension acquires form through a plane. Being indefinite and limitless in itself, when it takes on a surface and a limit (which is a kind of form) it becomes defined and acquires form.

> **209b9** For whenever the limit and the properties of a sphere are removed, nothing is left but matter.

The limit, that is: the surface and the spherical form; the properties, that is: the qualities capable of change – colour, weight etc. For when these have been removed, what is left is an indefinite and limitless mass, which is matter. Matter, that is: either prime matter or, even better, quantified matter, i.e. matter which is three-dimensionally extended, which in itself is indefinite and formless, for it is not identical to any determined quantity, like the two-cubits-long or three-cubits-long, nor is it identical to anything that exhibits form.[68] Hence it receives ever different sizes and shapes. There are even some who have thought that this is prime matter.[69]

> **209b11** This is why Plato also says in the *Timaeus* that matter and space are the same.

For this reason, he says, i.e. because place is like matter, Plato too says in the *Timaeus* that matter is the same as space; and place is space; so he says that matter is the same as place. He calls matter the participant, as I have already said, because it participates in the Forms, and participates only, while in no way being participated in. We have already said in what sense he also calls it 'space', namely in an analogical way, just as Aristotle himself calls the soul the place of the psychic forms and the intellect the place of the intelligible forms.

> **209b13** But there {sc. in the *Timaeus*, he} speaks in another way about the participant than in his so-called unwritten doctrines.

That is: he gives a different name to matter in the *Timaeus* from that **in his unwritten doctrines**, i.e. in his unwritten lectures. For in the unwritten lectures he called matter 'great-and-small', as Aristotle has indicated in what precedes, and we have said why matter is great and small. In the *Timaeus*, however, he calls matter **the participant** in virtue of the fact that it participates in the Forms. It was Aristotle himself who wrote down the unwritten lectures.[70]

> **209b15** But nevertheless he proclaimed that place and space are the same.

Even if, he says, Plato uses different denominations for matter, he nonetheless says that place and matter are the same. For if he calls matter space, and space is nothing other than place, then it is clear that place is the same as matter. Whereas all others say that place exists, but none of them what it is, Plato alone has left an account of what place is, namely matter. But we have said that he did not identify matter with the kind of place which we are discussing now, but called it thus by analogy, because just as place is a receptacle for bodies, just so is matter a receptacle for forms.

> **209b20** For {the question} anyway requires the keenest examination, and apart from each other it is not easy to gain knowledge of them {sc. matter and form}.

Along these lines, he says, the difficulty of the investigation of place is made plain. For regardless of whether place is matter or rather
522,1 form, the theory about it is difficult. For it is linked with the first principles, the contemplation of which in itself is the summit of philosophy and nothing simple. But also, because we here attempt to get to know in separation from each other things that cannot be known separately. For matter and physical form belong to the things that are relative: matter is said with respect to form (for it is the matter of a form), and similarly the physical form is the form of matter.

> **209b21** But that indeed it is impossible {for place to be} either of these.

Either of these: i.e. form or matter. From here onwards, then, he sets up the proof, through several arguments, that place cannot be either matter or form.

> **209b22** For matter and form are not separated from the object.

The first proof. Matter and form are not separated from the object; place is separated from the object; therefore place is neither matter nor form. The figure is the second.[71] He does not further support the major premiss, because it is evident, but he does further support the minor premiss, i.e. that place is separate from the object. For in that in which there was formerly air, when the air has moved out, something else comes in, the bodies replacing each other. It follows that place is separated from the object.

> **209b27** So place is neither a part nor a state, but is separable from each.

A **part** is for instance a hand or a foot or a portion of homoiomerous substances;[72] by a **state** he here means either the substantial form,

or simply any form, including the accidental form. For the accidental properties, whether separable or inseparable, move along with the object; but place is immobile and does not travel along with the object. Similarly a part is not separated from the object either, as long as it is a part. For when separated it immediately ceases to be {a part}. It follows that place in no way belongs to the object. But it is also possible to take 'part' in the sense of 'element' (like form and matter), which is perhaps better, so that the conclusion **so place is neither a part nor a state** will stand for 'so it is clear that place is neither
matter nor form'. But why do I just mention matter and form? After 523,1
all, place is not a state either, i.e. it is not even an accidental property of the object, since place is separated from the object when it is moved, but the accidental properties move along with it. For even the accidental properties that are separable move along with the object as it changes place, yet place does not move along with it, but is separated.

> **209b28** For indeed place appears to be some such thing as a vessel.

He further supports the idea that place is separable from the object. For, he says, place differs in no way from a vessel. After all, a vessel is a portable place. So just as the object is separated from its vessel (for evidently the vessel is different from the object {in it}), so it is also separated from its place.

> **209b30** On the one hand, insofar as {place} is separable from the object it is not form, and on the other hand, insofar as it contains, it is different from matter.

From here we get the second argument, which is given in two parts: it shows on the same grounds as the preceding argument that place is not form, and it shows that place is not matter either through the fact that matter is contained whereas place contains.

> **209b32** That which is somewhere is always thought both to be something itself and to have something other outside it.

The third argument. Place is outside the object, which is something different besides it. Matter and form are not outside the object, nor something different from that of which they are the form or the matter (for the compound is nothing other than the combination of matter and form). Therefore place is neither matter nor form.

> **209b33** It ought indeed to be asked of Plato, if one must speak 524,1

by way of a digression, on account of what the Forms and numbers are not in place, if place really is the participant.

He criticises Plato for the fact that if he says that matter is place, the result will be that numbers and Forms too are in a place. (Forms and numbers are here juxtaposed as being the same.) For if matter is the participant of the Forms, and matter is place, then the Forms will be in a place. Yet he does not think that the Forms are in place either, as Aristotle has said earlier:[73] For Plato too wants them to be separate. Hence either Plato was wrong to call matter place, or he is forced to say that Forms are in place, which is not his view. But we will reply to Aristotle, first, that Plato did not say that matter is the place of separable Forms, which he indeed claims not to be in a place, nor that it is the participant of these, but rather of physical forms.[74] And, secondly, we have said that in speaking about place, too, he was not speaking about the kind of place we are now discussing, but by analogy.[75] And we have also said on many occasions in what sense he called the Forms numbers, viz. in virtue of their defining and limiting capacity.

He adds **if one must speak by way of a digression**, because proving that matter is not place was something that follows from the discussion at hand, whereas saying that if matter is place the Forms will be in place – this was not something relevant to the present inquiry, whether the Forms are actually in place or not. Yet we can in fact take this too as something contributing to the argument at hand. For if it is an absurdity that the Forms should be in place, it is clear that matter cannot be place.

210a2 Again, how could {a thing} move to its own place?

The fourth argument. From here the argument is back on track and connects with what precedes. For what was in between was a digression directed at Plato. If then, he says, matter and form were place, nothing would move towards its proper place. For nothing moves towards its form and its matter; after all, it is already in this. Yet everything does move towards its own place: a clod of earth downwards, fire upwards. Therefore matter and form are not place. If you frame this same argument in the form of a categorical syllogism, you
525,1 will have the fifth argument: all things move towards their proper place; nothing moves towards its matter and form; therefore place is neither matter nor form.

210a3 For it is impossible that that of which there is no motion, and no up and down either, would be place.

This is the sixth argument. Its construction proceeds together with that of the preceding argument. The argument is as follows. In place there is 'up' and 'down'; in matter and form there is neither 'up' nor 'down'; therefore place is neither matter nor form. The words **for it is impossible that that of which there is no motion** serve the construction of both arguments. For in that in which there is no motion there will neither be up and down nor will any thing move towards it.[76] But in matter and form no motion exists. It follows that neither does anything move towards these, nor do they exhibit 'up' and 'down'. **So place will have to be sought in those things** in which there is evidently motion and which do exhibit 'up' and 'down'.

> **210a5** If its place is in the thing (for it must be, if it is form and matter), place will be in place.

The seventh argument. If form and matter are place, and if form and matter are in the object, then place as well is in the object. So when the object, i.e. that made up of matter and form, changes place and from one place comes to occupy another place, evidently both the matter and the form move along with it. For where the compound is, there necessarily the simple things that are its components are as well. So when the whole comes to be in a place, both the matter and the form will necessarily do the same. But these are its place. Therefore its place will come to be in a place. **So there will be a place of a place**.

> **210a9** Again, whenever water comes to be out of air, the place 526.1
> has perished; for the body that has come to be is not in the same place.

The eighth argument. If form is place, he says, then whenever water has perished and become air, evidently its place will have vanished because its form has perished. Yet they are at a loss to say what this perishing of place actually is. For they could not say that place hasn't perished but is in the body that has come to be from the water, i.e. the air. For that is not in the same place: after all, the form of air is a different one. So if the form of air is different from the form of water, and form is place, then the air is not in the same place as the water.

Chapter 3

> **210a14** After these considerations we must undertake to enumerate in how many ways it is said that one thing is in another.[77]

After showing first that place exists and wishing next to give the what-it-is, he turned the argument around towards the contrary conclusion, and showed that if we wish to give the what-it-is of place, place runs the risk of turning out not to exist at all. Having thus argued on both sides, and since he had said at the beginning of his account that we cannot extract the same conclusion from the {various} properties of place, he next showed on the basis of which properties it would seem to be matter, and on the basis of which it would seem to be form. He then established through various arguments that it cannot be either matter or form. Having said this, he next goes on to show what place is, and to set out what, in his view, are its properties. For once this has been shown – i.e. what place is – all the puzzles are solved that would lead us to think that place does not exist at all, and in particular those that arose from a wrong conception of the what-it-is of place, such as those assuming place to be three-dimensionally extended.[78] There are some puzzles that are also wrong-headed in their own right,[79] like the one claiming that everything that exists is in a place, or that everything that exists is either an element or composed of elements.[80] What he eventually wants, as I have said, is to give the what-it-is of place, but since the things in place are more evident than place itself, and since things in place are in something, he wants to enumerate the various senses of 'in something', and to state in how many ways one can say that something is 'in' something.

And he mentions as many ways in which a thing can be said to be 'in' something, as are commonly explained by his commentators in their commentary on his *Categories*,[81] except that he does not list all the senses of 'in' something, but only nine of them. He omits two of them, being in time and being in a subject. At the same time he also inquires whether anything can be in itself,[82] either in one of these senses of being 'in' something or in some other way, and it is clear that he is talking about physical objects, which is what his whole account is all about.

So he inquires whether there is something that is in itself, or whether all things are necessarily either nowhere or in something else. He says that since it is possible to be something primarily, but also in virtue of something else – for Socrates is said to be white not in the primary sense, but in virtue of something else, namely in virtue of the fact that a part of him, his body, is white; but his body is not white in the primary sense either, but its surface is white in the primary sense, and the body is white on account of this, and Socrates on account of that: so Socrates is white in virtue of something else, not primarily. Similarly, Socrates is said to be shod, not primarily, but in virtue of something else, namely in virtue of the fact that a part of him, his foot, is shod. Similarly he is said to see, in virtue of a part of him, his eyes – since, then, it is possible for things to be something primarily, but also through a part, or more generally, in virtue of something else, for a thing to be in itself would also be possible either primarily or in virtue of a part. Now it is impossible for a thing to be in itself in the primary sense; but it is possible for it to be so in virtue of a part of it or of something else, if one part of the thing said to be in itself is something like a place and that which contains, and another what is in this. Since one of the parts of the whole is in the other, therefore the whole thing may be said to be in itself. For example, a jar of wine may be said to be in itself;[83] not primarily, nor in virtue of it being in itself as a whole, but of the fact that one of its parts is in another part. Just as we say that man is self-moving, not because the whole of him moves and the whole of him is moved but because a part of him moves whereas another part is moved (the soul being what moves, the body what is moved), so you may also say that the man is in himself because one part of him is in another part – his soul in his body – and that the jar of wine is in itself, because one part of it is in another part, viz. the wine in the jar. Only in this way can a thing be in itself, in virtue of a part, as I have said, and by an improper use of the term. Common speech, as we already said, knows this usage of calling things as a whole after a part of them too, as when we say that Socrates is shod or sees, or that Socrates is knowledgeable. For it is in virtue of the fact that his soul is knowledgeable, or his eye sees, or his foot is shod, that the whole of him is named after a part. In this way then, here too, the whole can be said to be in itself in virtue of the fact that one of the parts of the whole is in something, though not in something else apart from this whole. Since also each of the body parts of a living being – hands, feet and the rest – is in the whole, this whole may also be said to be in itself not primarily, as I have said, but in virtue of something else and not in the proper sense. In this way a jar of wine too may be said to be in itself as in a place, through the fact that part of it is in the rest of it as in a place. In virtue of the fact that the whole is nothing other than the place plus the emplaced, in that sense the whole is said to be in itself. Only in this way can a thing be in itself; in the primary sense it cannot be in itself, neither in its own right nor incidentally. And he shows this first through the method of sampling of cases and then through the method of proof.

Alexander adds still other senses of being 'in' something. As one of them he mentions that of things being together, i.e. things being in touch with their surfaces. For these may be said to be 'in' each other.[84] But it is clear that things that are together are always said to be {together} as either in time or in place. Now things that touch each other are said to be in one another as in place. For the surfaces with which they touch are part of the place that surrounds them.[85] Hence, this is not a new sense of being 'in' something next to the ones enumerated. He adds yet another sense, which is not unsubtle. For, he says, a thing is said to be 'in' something as a subject in what is accidental, such as when we say that this or that person is in fine circumstances, or I am in a bad situation: then it is our affairs that are in this state. In this sense we say that a subject is 'in' an accident.

> **210a18** In yet another way, as the genus is in the form, and generally, part of the form is in the definition of the form.

It need not surprise us if the genus is said to be in the form, i.e. the greater in the lesser.[86] For genera are both common natures that belong to various things and are predicated of various objects, and also become parts of forms, when they are considered as things that are already present in the forms{, i.e. species}. Hence, according to the perspective taken, here the same things are at once more general and more specific: when they are considered as being not yet connected with the appropriate differentiae they are more general, but when they are thus interwoven and so come to constitute the forms, then they become parts of the forms. For of the definition of man, i.e. rational mortal living being, 'living being', while being the genus, is a part. This at least is why Aristotle himself too, having said **in yet another way, as the genus is in the form**, in order to show that he uses **genus** not in the sense of that which is predicated of various objects that differ in form, as part of their essence, but in the sense of that which is already connected with its differentiae and has become part of the forms – this is why he has added **and generally, part of the form is in the definition of the form**. For 'living being', while being part of {the form} human, is included in its definition.

> **210a21-2** Further, as the affairs of the Greeks are in {the hands of} the king {of Persia}, and generally {as things are} in the {power of the} first thing productive of change.

For the ruler is the active cause of political affairs. Thus Homer too says: 'It is in the lap of Zeus', and this is common parlance too; we often say that life and death lie 'in' a ruler's hands.

> **210a24** And in the most fundamental sense of all, as {a thing is} in a vessel and, generally, in a place.

If you take 'as in a vessel' and 'as in a place' as two different items, the whole list consists of nine items; but if you take 'as in a place' as the more general item and on the understanding that the vessel is as it were a part or species of place, there are eight in all, except that either way there are two items that have been left out: being in something as in time, and being in something as in a subject. It is natural that he said that, of all the senses of being 'in' something, the **most fundamental** one is the sense of being 'in' **as in a place**. For in these cases the distinction between that in which and that which is in it is a clear one.[87]

210a25 One might wonder whether it is possible for something to be in itself, or whether nothing can, and everything is either nowhere or in something else.

Having enumerated the senses of the phrase 'in something', he naturally inquires whether a thing can also be said to be in itself – and if it can, whether this falls under one of the senses enumerated or whether a thing is said to be in itself in yet another sense; or whether, he says, no thing can be said to be in itself but everything in existence is either in something else or is nowhere. For there are no other options: a thing is either in itself, or in something else, or nowhere at all.

210a26 This {sc. 'in itself'} is in two ways: either in respect of itself or in respect of something else. 530,1

In respect of itself here needs to be taken as equivalent to 'primarily'.[88] For one needs to know that 'in respect of itself' is opposed to 'incidentally', and 'in respect of something else' to 'primarily'.[89] What is said to be walking 'in respect of itself' is Socrates, but incidentally the white or the bald. 'Primarily' can be used both in the case of what is in respect of itself and in the case of what is incidentally,[90] and so can 'in respect of something else'. For example, whenever I say that Socrates is walking, I have said he is so in respect of himself and in the primary sense, and I mean by Socrates the ensouled living being: for he does not walk in respect of anything other than in so far as he is a living being.[91] But whenever a ship is moving and I say that its sailor is moving, then I have pronounced moving incidentally as well as in the primary sense: incidentally, because that which moves in respect of itself is the ship, and primarily, because the sailor is not moving in respect of anything else, but *qua* sailor. When someone moves the door with his stick, he is said to move the door in respect of himself *and* in respect of, or through, something else, but when he moves the door with his stick and I say 'the white {thing here} moves the door' I have expressed a moving that is incidental and in respect of something else. It is not only when we set something in motion by means of an instrument that we are said to move it in respect of something else, and not primarily (for the carpenter does not move in the primary sense the wood, but primarily the axe, and through this the wood), but also when we do something with a part. For example, when Socrates moves his finger or his foot in going to sleep, Socrates is said to be moving, yet not to be moving primarily but in respect of something else: through moving his foot or his finger.

Hence, one would not even say that Socrates is seeing primarily, if what is said to be 'primarily' is that which is not predicated in virtue of some other aspect of the rest of the whole thing but in virtue of itself. So if Socrates is the whole of the living being, and this living being is not seeing through the whole of itself but through a part, then 'seeing' said of Socrates is a seeing 'in respect of itself' but through something else, that is: through a part and not *qua* being Socrates.[92] For he is Socrates not through a part, but in so far as he is the whole thing, yet he is seeing not in so far as he is the whole thing but in so far as he has sight. But since people often describe the opposite of 'primarily' as 'in respect of something else' but often also as 'through something else', as if these two expressions both meant the same thing, it makes more sense, if we want to use the terms in their more proper senses, to say 'through something else' whenever we effect something through a medium which is not natural (e.g. when we move the door by means of a stick), and 'in respect of something else' whenever we effect something through a part of us.
531,1 So much for these things.

Aristotle says here that a thing's being in itself is either in respect of itself or in respect of something else, using **in respect of itself** as equivalent to 'primarily'. Having first shown that something can be in itself in respect of something else, he then shows that it is impossible for something to be in itself primarily – and 'primarily' neither in the sense of 'in respect of itself', nor in the sense of 'incidentally'.[93]

> **210a27** For whenever that in which something is and that which is in it are parts of a whole, the whole will be said to be in itself.

In what sense a thing will be in itself in respect of something else is what he here shows, namely: in virtue of a part. For it is in virtue of the fact that one part of a whole is in another that the whole can be said to be in itself. And saying this is non-proper usage, seeing that everyday language too is capable of labelling wholes after a part, e.g. by saying that Socrates is shod, and that he is white, or knowledgeable; and yet it is a part that is shod, the foot, or that is white, his surface, or that is knowledgeable, his soul. In this way a jar of wine may also be said to be in itself, because a part of the whole is in another part. But do not demand that the situation fits this example in all respects. In the one case the foot is shod and so the whole is said to be shod, all of it being named by the name-label for the part, but with the jar of wine it is different: here it is not because the part is in itself that the whole is said to be in itself. However, let the example be useful to the extent that things are often named after a part. And since in the case of the jar of wine, too, one of its parts is in a place, but not in a place that is different from or outside the whole – no, both the place and that which is emplaced are within the whole – for this reason the whole *and* the place *and* the emplaced may be said to be 'in itself'. And since a part of it is the place and another part the emplaced, if the place and the emplaced are the same thing, there will be something in itself as in a place. As an even more congenial example you have the self-mover: for the jar of wine is said to be 'in itself' in just the way that the animal is said to be self-moving.

> **210a32** For *what* {a thing is} and what {it is} *in*, both are parts of the same thing. 532,1

By 'what' he means what is in a place, i.e. the wine, and by 'what {it is} in' he means the place, which was the jar. Now the wine and the jar are parts of the whole, i.e. of the jar of wine. When taken separately, he says, none of these parts can be said to be in itself, but the whole can, as has been said.

> **210a33** But it is impossible {for a thing} to be in itself primarily. {It is} just as white is in the body, <for the surface is in the body,> and {as} knowledge is in the soul.

Having shown in what sense a thing can be in itself, namely in respect of something else, he now says that **it cannot {be** said to be **in itself} primarily**, neither in respect of itself nor incidentally. First he shows that it cannot be so in respect of itself, then that it cannot be so incidentally either. Since he first shows that it cannot be so in respect of itself, **primarily** should here be taken as 'in respect of itself primarily'. And although he has said that **it is impossible** {for something to be in itself} **primarily**, the examples he has adduced do not apply to this any longer but to what he had said just before, viz. that some things are said of things in virtue of a part.[94] For having said 'in this way a thing can be said to be in itself', i.e. in virtue of a part, he has now again adduced examples of what is said in virtue of a part. For it is in virtue of the fact that the surface that contains the whiteness is in the body, that the whiteness too is said to be in the body.

> **210b1** The labels are in respect of these things, being parts, at least in the case of a man.

The labels, he says, 'white' and 'knowledge' are predicated of the whole not primarily, but in virtue of its surface and of its soul, which are parts not in themselves but as they are in a human. For taken in themselves the surface and the soul are wholes, but considered as existing in a human they are parts (and by 'human' I mean the

25 compound of soul and body). So if the name-labels apply to the whole in virtue of these things, and these things are parts, the whole is being labelled after a part.

533,1 **210b2** The jar and the wine, when they are apart, are not parts, but when together they are; this is why whenever they are {considered} *qua* parts, a thing will be in itself.

For if the part is the part of a whole, and part and whole belong to the things that are relative to each other, then whenever jar and wine are separated, they will not be parts, being unrelated to each other.
5 But when they come together and become parts of each other, then the whole is said to be in itself in virtue of a part, just as a man is said to be white in virtue of his body, and his body in virtue of its surface. Once again, he has come up with the same examples of the same thing.

10 **210b6** And these, surface and white, are different in form and have a different nature and capacity.

Since the whiteness is in the surface in the primary sense, and in order to avoid that someone takes the whiteness to be the same thing as the surface, and takes it that any thing is in itself in this way, he adds that these two things are different from each other in both form and capacity. For the surface is always a subject, not only for white-
15 ness, but also for blackness and various other qualities, and the whiteness can only exist in a subject, and it is impossible for it to exist together with its contrary. Now if these things differ from each other, then one is in the other. He probably uses **nature** because the one thing is a quality and a colour whereas the other, surface, is a limit of the body and of quantity; and he probably uses **capacity** because the one thing is a subject whereas the other belongs to the things that are in a subject.

20 **210b8** So running through the examples[95] we see nothing that is in itself on any of the definitions made, and also by reasoning it is clear that this is impossible.[96]

Having shown in what sense it is possible for something to be in itself, namely that it is possible for something to be so in respect of something else, he now wants to show that it is not possible for something to be in itself in the primary sense either in respect of itself or incidentally; and he first shows that it is not possible for a
25 thing to be in itself in respect of itself. He does so both through examples and through reasoning, and first through examples.[97] For
534,1 if there are as many senses of being in something as we have

enumerated, and if in none of these senses a thing can be in itself, then it seems that nothing at all can be in itself. For all the senses of being in something pertain to one thing being in another. For not even when we say that the whole is in its parts do we say that it is in itself. For the whole is one thing, and its parts are something else. The former is one and the same (being a stationary condition of the connection between the parts, and a form which supervenes on their connection), the latter are plural and different.

But he also shows this through reasoning, as follows. The definition of what is receptive {of something} and the definition of what is in it, he says, are different. For place is defined differently from what is in place, and in general that in which {something is} is defined differently from what is in it. If, then, we suppose something is in itself, it will admit different or rather opposite definitions. To put it more precisely, a thing itself will in respect of the same thing be different and opposite things. For it is not impossible for the same thing to admit two definitions according to two different distinctive properties (for example man can be defined as a rational mortal living being or as a living being that walks upright), but for it to admit contrary accounts, at the same time and in respect of the same thing, is impossible. For place and that which is in place belong to the things that are relative, and relatives are, at the least, pairs of opposed items, having contrary definitional accounts. Now if a thing is to be in itself, for example if a jar of wine is to be in itself, it will admit contrary definitions, and since the jar is a place and the wine is in a place, the same thing will admit the definition of place and of that which is in a place. The jar will then be at once jar and wine, and the wine will be at once wine and a jar. And that is ridiculous. Next, in an attempt to support this view, in order to be able to refute it by going round and taking it in the rear, he says that, even if we concede the main outlines in virtue of the fact that a part is in a part and each of them is and is said to be in the other, so that the jar is in fact both jar and wine, and the wine is both wine and jar – even so nothing will be in itself, but each of them will be a container not *qua* being wine, but *qua* being a jar, and will be contained not *qua* being a jar, but *qua* being wine. Hence, not even in this way will anything be in itself, but one thing will be in another.

Having thus shown that nothing is in itself in respect of itself, he next goes on to show that nothing can be in itself incidentally either.
It will probably be thought that if a thing can be in itself in respect 535,1
of something else, it can also be in itself incidentally. In order to be able to follow the proof, we must first get our conception clear of what it is that is 'incidental', because Aristotle is both very brief and unclear in the way he conducts the proof.[98] A thing is said to have a property incidentally, if it is itself an accidental property of something else, and if that property which it is said to have incidentally

belongs in respect of itself to that of which it is an accidental property.[99] For example: what is white is said to be in motion incidentally, because motion belongs to Socrates in respect of himself, and because whiteness too belongs to Socrates as an accidental property. For the reason why motion belongs to the white incidentally is because it belongs to that of which the white is an accident, i.e. to Socrates, in respect of himself. Now if the jar is in itself incidentally, then something else must be in it in respect of itself, for example the wine, and the jar must then be an accidental property of the wine, in order that, by the fact that the wine belongs to the jar in virtue of itself, that which is an accidental property of the wine, i.e. the jar, will be in itself incidentally. There is no other way in which the jar can be in itself incidentally. And no one should demand that the explanation we have given of the incidental fits directly the case of the jar, for here things work in a different way: here one thing does not incidentally belong to another as a property, as it did in our explanation, but a thing itself is in itself. We have yet to find out how we can accommodate the incidental in this case as well, given the fact that there is no other way the concept of the incidental can possibly be saved than in the way we have explained. Now, if the jar belongs to the wine as an accidental property, it is clear that it has pervaded its whole substance (for if it has not pervaded it but merely touches it at its limits, it no longer belongs to it as a property). But if it does completely pervade it, then a body pervades another body. Apart from that, he says, the argument will come round to the point we started from. For the jar will end up being in itself in respect of itself. For what is the fundamental difference between saying that the jar is in itself, and saying that it is a property of the wine that is in the jar in respect of itself? For *via* the wine it will end up being in itself, and it will receive the two different definitions, that of place and that of what is in a place.[100] We may conclude that a thing cannot be in itself incidentally either.

Having said that, he also solves Zeno's puzzle on the basis of the above overview of different ways of being 'in something'. The latter came up with the puzzle that, if place belongs to the things that exist, and if everything that exists is in something, then place too will be in something; hence, place will be in a place, and so on *ad infinitum*. It is not difficult, he says, to solve this puzzle on the basis of the senses of being in something that we have enumerated. For even if it is not true that everything that exists is somewhere, <still it is true that everything that exists is in something> according to one of the senses of being in something (for it is necessary that every physical object is in something but not in a place).[101] Hence, if place too belongs to the things that exist, it will for sure be in something. But given that being in something is said in many ways, place too will be in something, though not in a place. Rather, just as states and affections are in something (for example in a body) not as in a place but as in a subject, so also place in the proper sense – I mean the limit of the container – will be in something (for it is in a body), yet not as in a place, but as a limit in that which is limited, which amounts to the same thing as: as in a subject.

On the basis of the things said he also, once more, proves something else, which he has already proved in a number of ways. For if the proof is established that this is the reason why a thing cannot be in itself, viz. that the nature of what is apt to receive is different from the nature of what is received and that it is not possible for one and the same thing to receive different definitions – then on these grounds as well it becomes clear that place cannot be either matter or form. For what is in place *is* different from place {itself}, but the matter and the form are not different from the thing that is 'in' them. For they are the principles that make up its complete nature. Therefore place is different from matter and form.

> **210b8** So neither investigating inductively

– that is: through a sampling of cases and looking at individual cases. **On any of the definitions**, by which we have defined being 'in' a thing. Not only, he says, from these, but also on the basis of reason itself it is plainly impossible that a thing should be in itself.

> **210b10** For it would require that each thing would be both, for example a jar would have to be both a jar and wine, and the wine would have to be both wine and a jar.

If we are going to say, he says, that the jar of wine is in itself, in respect of itself and primarily, as a whole in a whole (for that is what we mean by 'in respect of itself and primarily), then the whole thing as a whole must necessarily receive the definitions both of the place and of the emplaced, or rather: it has to *be* both. The jar figures as a place, and what is emplaced is the wine, or simply any liquid. This means that when the jar of wine is in itself as a whole, then the jar too – that is: the ceramic vessel – will not only be a jar, but also wine, and the wine not only wine but also a jar. Hence, if all this is absurd, it is impossible for a thing to be in itself in respect of itself and primarily.

> **210b13** So, however much they may be in each other, the jar will contain the wine not in so far as it itself is wine, but insofar as the other item is.[102]

As we said in our introduction, in these words he speaks unclearly and in riddling fashion. For having brought the argument to the absurd conclusion that the jar is also wine, and that the wine is not just wine but also a jar, he envisages that someone might say: 'But why is this absurd? For in virtue of the fact that the one is in the other, each of them may be called by the name of the combination; for there is one whole comprising both items'. He grants this argument, but shows that even so a thing is not in itself, but rather one thing is in another. For, he says, however fully one may grant that the jar is also wine, so that in this way it will be in itself, and similarly that the wine is also a jar, it is surely clear that the jar will not receive the wine in so far as it is wine, but in so far as the other item is wine. And similarly, the wine will be in the jar not in so far as it is itself a jar, but in so far as the other item is a jar. Hence, each of them will have the name of the whole, but not its reality.

> **210b16** It is clear that in respect of their being they are different; for the definition of the 'that in which' is different from the definition of what is in it.

A fitting conclusion at this point would be: 'it is clear that it is impossible for a thing to be in itself in respect of itself'. But this is not what he says; what he does offer as a conclusion is that on the basis of which he has shown that it is impossible for a thing to be in itself in respect of itself, viz. that by nature the place and that which is in place are different. For you will define differently that which is receptive, and that which is received. Hence, if things whose definitions are different are themselves different too, and if the definitions of the that-in-which and of what is in this are different, these too will be different things and other than each other.

538,1 > **210b18** Nor indeed is it possible incidentally; for {then} two things will be in the same thing: the jar will be in itself.

Here he briefly, and as a result also unclearly, shows that a thing is not in itself incidentally either. For, he says, two bodies will be in the same place and we will thus come round to the same argument that we also encountered in the case of {what is in itself} in respect of itself. For the jar will, once again, be in itself and will receive the definitions both of place and of what is in place. Two absurdities follow from the hypothesis, which he has briefly set out. If that which is of a receptive nature can be in itself incidentally, then what it is receptive of must be in it as well[103] – for example, if it is receptive of wine, then wine – and the receptive must be an accidental property of the wine (something Aristotle in his brevity has omitted to mention). For only thus can it be in itself incidentally: by being an accidental property of the wine, which in its turn is in the jar in respect of itself. Two absurdities follow from this: that a body will pervade another body, if indeed the jar is an accidental property of the wine, and, once again, the jar will be in itself in respect of itself,[104] which has been shown to be impossible. For the jar is in the wine; but the wine is in the jar; hence the jar is in the jar as well.

> **210b22** It is not difficult to solve Zeno's puzzle, namely that since place is an existing thing, it will be in something.[105]

If this was Zeno's puzzle – viz. that if place is in something (since every existent thing is in something), and if what is in something is in a place, then place is in a place – then its refutation is really simple. For 'in something' is said in many ways, not merely in the sense of 'in a place', so that nothing prevents place from being in something yet not being in a place. But if, as was described earlier, he did not construct the argument by claiming that place is in something and hence is in place, but by claiming that everything 539,1
that exists is in place, and that place belongs to the things that exist, and so it is itself in a place as well, it will not be refuted by what is being said now, but by what has been said there.[106] For we shall not grant the first premiss, that 'everything that exists is in place'.

> **210b24** For nothing prevents {place} in the primary sense from being in something else, however not in the way things are in that place itself.

He uses place **in the primary sense** instead of 'in the proper sense', and that is {place as} the limit of the surrounding body.[107] This he said, because the jug is also said to be the place of the wine, and the air, the earth and a number of other things are said to be our place. Things called place in that sense are also themselves in a place, but when we take place in the proper sense, it will be in something, for example in a body (for it is the limit of this), yet not as in a place but as a property of it.

> **210b27** But this is obvious.

From here, as I have said, he once again shows on the basis of what has been said, that place is neither matter nor form. Place, he says, is nothing of the things that are in place, besides being their place. Matter and form are principles of what consists in them, i.e. of the compound, and are nothing separate besides it. Therefore place is neither matter nor form.

Chapter 4

210b32 What place is, may become clear in the following way.[108]

From here on he wants to take the remaining step and to establish what place is, and he first sets out the characteristics that belong to place according to the common conception.[109] Next he will tell us whence in general we human beings have conceived the conception of place, and having given his explanation he will present some theorems that follow from it.[110] Then he will take the remaining step and produce his account of place in such a form that it will at the same time solve, as he himself calls it, the puzzles that have been brought up with respect to place, while it will also show that the characteristics that according to the common conception belong to place in respect of itself do indeed belong, and will make clear the cause of the difficulty that surrounds the theory of place.[111]

540,1 First, then, as I have said, he sets out the characteristics that belong to place according to the common conception, which are four in number. The first is that place contains what is in place (for this we all know in virtue of our common conception: that place contains what is in it, as the jar contains the wine – for a jar too is a kind of place) and that in containing it the place is not anything of the object that is contained. This latter characteristic Themistius affirms to be the second among the characteristics, so that, with the three others that follow, the sum total is five,[112] but I think it is right to present this combination – that place contains and that it is not anything of the object – as one whole, since a surface too is taken to 'contain' a body, and we say that the whole 'contains' its parts, yet place is not said to contain what is in place in the way that a whole contains a part. For the whole is indeed some feature of the part, but place contains without being anything of the thing contained – neither a limit of it, nor its wholeness, nor an affection, nor anything else. And Aristotle too makes it clear that he intends this claim to be taken as one whole, in so far as he formulates what follows as if he were making a new start, by saying **Next, that the primary place is neither smaller nor larger than that of which it is the place**.[113]

This, then, is the second characteristic belonging to place, that place is neither larger nor smaller than that which is in place – and place here is the immediately enclosing proper place of each thing. For if it is smaller, then it does not contain the body, and thus not every body – I am speaking of bodies in the sense of *parts* of the cosmos – is in a place.[114] And if it is larger, then it surely can receive yet another body, and so it will no longer be place in the primary sense. Another result will be that there is also void, which will shortly be shown to be impossible.[115]

The third characteristic is that place should **not be left behind**

by each object and that it should **be separable**. Two readings are current: either 'should be left behind' or, with a negation, 'should not be left behind'.[116] If the reading is 'should be left behind', as indeed the majority of the manuscripts have it, then he is speaking of the individual place which immediately contains the individual object, which he claims is left by what is in place and is separated from it (for **to be separated** is an explanatory addition to **to be left behind**). If the reading is 'should **not be left behind**', he would not be talking about the individual place, but about place in general. For place in general is not left without bodies. For sure, a body is in some particular place; for even if it leaves this place behind, yet it will at any rate be in some other place, and it will never be left behind by place *tout court*. **Being separable**, according to this reading, is equivalent to being not any feature of the object, but on the contrary being outside its essence and not contributing anything to this.

The fourth characteristic is that **place as a whole has 'above' and 'below'** – not simply each individual place (for each individual place is either above or below), but in place as a whole there is above and below.[117] Aristotle's exegetes add a fifth characteristic as well: that place is immobile – something he himself also adds a little later.[118]

These being the characteristics that thus apply to place according to our common conception, what is required is to give an account of place such that it solves the puzzles that are being brought up against it, that it will safeguard the characteristics that are held to apply to it as indeed applying to it, and that it will make clear the cause of the difficulty of the theory of place.[119]

He next investigates how human beings arrived at the conception of place, and he affirms that they got it from nothing other than change with regard to place.[120] For if things were immobile, we would never have arrived at the notion that place is one thing and that which is in place another (for the whole would have been as if continuous and one). As things are, however, by seeing that things are moving and are now here, now there, we have arrived at the notion that that which is able to receive, which is place, is one thing, and that which comes to be in it is another thing. This, he says, is why we think of the heavenly bodies in particular that they are in place, because it is also these in particular in which we witness change with regard to place.[121] Having said that change with regard to place is what has led us to the conception of place, he lists its various kinds, in order to be able to show more clearly for each of these kinds, that these or these are the causes of change with regard to place. There are, he says, three kinds of change with regard to place: locomotion, increase and decrease. For also things that increase or decrease exchange one place for another, even if not as a whole, then at least with respect to their parts. But as a whole, too, even if they do not exchange one place for another, yet they do occupy, in their entirety, now a larger place (when they increase), now a smaller place (when they decrease).[122]

Among things that exhibit locomotion, he says, some move in respect of themselves, others incidentally, and among the things 542,1 moving incidentally, some are such as to be able to move in respect of themselves as well, like the sailor, he says, **and the nail in the ship** and the parts of the whole. For both the sailor and the nail in the ship move incidentally when the ship is moving, and parts too move incidentally when the whole is moving. But all these things, while moving incidentally, are also capable of moving in respect of themselves. After all, the parts, when cut off from the whole (for example a hand or a foot) can move in respect of themselves. For even if it is no longer a part when it has been cut off, still it is able to move in respect of itself as a substance. Accidents, however, such as whiteness, or being snub-nosed, and things like that, while they do move incidentally when the man moves {whose accidents they are}, are unable to move in respect of themselves, since they cannot be separated from that in which they inhere. He has added all this, after having said that it is locomotion which has led us to the conception of place, because he wanted to show that it is not things in general moving in any way whatsoever that contribute to the conception of place, but only the things that move in respect of themselves, and that the things that move incidentally, since they do not move in the proper sense, do neither require place nor contribute to its conception.

Having said this, he distinguishes between things that are in a whole and things that are in a place, since many characteristics that apply to primary place, which is also place in the proper sense, i.e. that which immediately contains each thing, also apply to the whole. For the whole, too, contains its parts, and is neither larger than these parts {taken together} nor smaller. Furthermore, just as things each move towards their own places, so they move towards their own wholeness.[123] Before this, he once again says what place properly speaking is, that it is not as if, because we are said to be in the cosmos, the cosmos were our place, but that we are in the air, and that we are said to be in this because we are in this part of it here that immediately contains each of us; and so this is our place properly speaking. This, then, being place in the proper sense, being in a whole differs from being in a place in so far as that which is in a whole is continuous with the whole in which it is said to be, whereas that which is in a place is not continuous with the place, but is merely in contact with it (for the limit of the place differs from the limit of what is in the place), and in so far as that which is in a whole does not move *in* the whole but moves *with* the whole, whereas what is in a place, whether this place (as e.g. a jar is a place) is moving or immobile, can move in respect of itself in its place. For the things that rotate move in their own place, as do liquids in their vessels, such as water in a pot or a jar.

211a1-2 Also, that a thing's primary place is neither larger nor smaller.

And how, if place contains, will it be neither larger nor smaller? After all what contains is larger than what is contained. Now I say that if you were to claim that the whole body of air contains and is a place, 543,1
the container would in fact be larger, but if place is not this, the whole of the air, but the limit of it that immediately contains and that is place in the most proper sense, then this will be neither larger nor smaller. For the surfaces, that of the place and that of the thing in place, fit onto each other.

211a4 and by nature move to and remain in their proper places.

He has not adduced this as yet another characteristic of place, but as something following from **every place has 'above' and 'below'**, or indeed establishing this.[124] After all, 'above' and 'below' belong to places in the sense that each thing by nature moves to its own place, and that when it is in a place contrary to its nature, it will not stay in it but by a natural impulse move towards its proper place, but having arrived at its natural place it will remain in this and not be removed from it other than by force.

211a6 With these things established, the rest must be investigated.

That is: the things following from these.

211a7 We should try to conduct our investigation in such a way that the 'what it is' {of place} is established so that the puzzles are solved, etc.

In these words he gives us a most useful criterion as to how we should formulate our theories about things. It is required, he says, that the account we provide about place must be such as to also solve by and through itself every puzzle that is raised with respect to the subject at issue, to preserve within itself all the characteristics that belong to the subject according to our common conception, and also to show by and through itself what is the cause of the difficulty surrounding the discovery of the {nature of the} subject, and what it is that has given rise to the puzzles concerning it. The wording, which is syntactically a little difficult, should be taken as follows: **we should try to conduct our investigation in such a way that the 'what it is' is**

provided in such a way, etc., so the words **in such a way** belong to both clauses.

> **211a12** First it is necessary to understand that place would not be inquired into if there were no change with respect to place.

With these words he investigates how we have arrived at a conception of place to begin with. He claims that this happened based on {the phenomenon of} change with respect to place.

544,1 **211a14** Of this, one kind is locomotion, the other increase and decrease.

Of this: not of change in general, but of change with respect to place.[125] That increase and decrease too are forms of change with respect to place, he establishes on the basis of the fact that things that increase do not occupy the same place, but at one time a larger one, at another a smaller one. Now given that a thing, to the extent that it increases or decreases in size, at all events changes its place, to this extent these changes are surely changes with respect to place.

> **211a17** That which moves does so in actual fact in respect of itself, or incidentally.

Since he has said that it is from things changing with respect to place that we have formed the conception of place, and having explained what change with respect to place is, he also wants to provide a division of the things that move with respect to place, since not all things that are in any way said to change with respect to place contribute to the conception of place, and in order to show which ones do and which ones do not contribute. He says that among the things that change with respect to place some do so in respect of themselves, others incidentally. Now, none of the things moving incidentally contribute to the conception of place; only the things that move in respect of themselves do, even if some of the things that move incidentally are capable of moving in respect of themselves as well, like the sailor and **the nail in the ship** and the parts of a whole. Nevertheless, as long as they move along with the whole, they do not contribute anything to our conception, for they move incidentally. Since they do not move in the proper sense, they do not need a place either. For they are not in a place in respect of themselves. This is why they do not lead us towards a conception of place either.

> **211a23** Since we say that we are in the heavens as in a place because we are in air, which in turn is in the heavens.

Here he wants to distinguish things that are in a place from things that are in a whole, and he starts out by showing, once again, what place in the proper sense actually is. Since he has in a general sense mentioned things that are in a whole – of which he claims that they are capable of moving in respect of themselves, but that nevertheless they move only incidentally as long as they are in the whole – he 545,1
wants to distinguish these from things that are in a place. For in this way it will be shown more clearly that they do not contribute to our conception of place.

> **211a27** For, if the air as a whole were our place, then the place of each thing and the thing itself would not be equal.

That it is not the air as a whole which is our place, but that limit of the air which immediately contains each thing, this he shows on the basis of the preconceived axiom about place, namely that it is equal to what is in it and neither larger nor smaller. If the whole of the air was a place, then the place of each thing will turn out not to be equal to the thing itself; but as things are, it is equal; therefore the whole air is not a place. But even if it is in fact said to be our place, it is named thus after a part, on account of the portion of it that immediately contains us.

> **211a28** And the primary place in which a thing is seems to be like this.

The primary place, he says, in which each individual thing is, **seems to be** equal to this thing, which is in it as in a place. He uses **seems** as equivalent to 'appears' and 'is'.

> **211a29** Whenever that which contains is not divided but continuous.

Having said what place in the primary sense is, he gives a further distinction between things in place and things in a whole, since in the case of a whole too the parts are contained by the whole and the whole is neither larger nor smaller than the parts {taken together}. So what is the difference? That the parts of the whole are not divided, but continuous with the whole, he says, but things that are in place in the primary sense are not continuous {with place} but are only in contact with it, while remaining divided from it.[126]

> **211a32** … {the extreme} …, that is neither a part of that which is in it nor larger than its extension but is equal; for the limits of things that are in contact coincide.[127]

The limit of that which contains, he says, is neither a part of what is within it, yet extraneous to it and not belonging to it, nor is it larger or smaller than it. In order to establish this too, he has adduced the
546,1 words 'for the limits of things that are in contact coincide', that is: things that are in contact do so with their limits together and in the same place, and limits of bodies are surfaces. Since the surfaces fit onto each other, the surface of the place will evidently fit onto the surface of the emplaced, and when these fit onto each other they become one, so that they are both in the same place.[128] But if they are in the same place, they are also equal. Therefore what contains is neither larger than what is contained nor smaller, but is equal to it.

> **211a34** And when it is continuous {with it}, it is not moved in it but with it, whereas when divided {from it} it moves in it.

Yet another distinction between things that are in a place, and things that are in a whole. The parts, he says, being continuous with the whole, do not move *in* the whole but *with* the whole (for when the whole moves, the part as well moves along with it, yet it does not move in the whole as in a place). Things that are in place, however, being divided from the place, do not move *with* their place but *in* their place, the way a ball, or the water in a jug, rotate within their place while this place itself is immobile. And, he says, even if the place, for example a vessel, moves, still the water does not move with it but in it. For, generally, it is not even the vessel that constitutes the primary place, but its hollow surface, which is in fact immobile. We would neither say that the hand moves *in* the whole (but rather that it moves with the whole), nor, that those things that are in place move *with* their place (but rather that they move in their place). Therefore, things that are in a place and things that are in a whole are not the same thing.

> **211b1** Again, whenever a thing is not divided it is said to be as a part in a whole, like sight in an eye or a hand in a body.

He appears to be repeating himself: he states, once again, the same things. But since above he has given the arguments without examples, he now repeats them in order to add the examples.[129]

> **211b5** So, by now it is clear from the foregoing what place is.[130]

In what follows from here on he wants to formulate what place is. Since, he says, there are four items of which place must be one, if it can be shown for three of them that it is not that one, then, necessarily, the one that remains is what place is. For on the basis of the
547,1 characteristics of place we are led to believe that it is either matter,

or form, or the extension between the limits of the container, for example the extension in between the hollow surface of the jar, or these limits of the container themselves, for example the hollow surface itself of the jar. In so far as place contains and defines what is in place and is equal to the latter's limit, it is thought to be form. For a thing's contours and its limit are its form, I mean the form that has to do with shape and outer appearance.[131] In so far as place remains one and the same while receiving ever-different bodies, matter appears to be place. For it is peculiar to matter to receive ever-different forms and essences. Again, in so far as place, being separate and different, receives ever-different bodies while remaining one and the same, it appears to be an extension that is different from the bodies that come to be in it. For we have not arrived at the conception that place is something from anything other than the mutual replacement of bodies. If, then, it receives these bodies that replace each other while itself remaining one and the same, and if the things which replace each other entering into it are three-dimensionally extended, then it is necessary that the space which receives them is so too; so that place too will be three-dimensionally extended. Again, in so far as the limit of the container is in contact with the thing that is contained and does not allow any extension in between except the body contained, and {in so far as} in the mutual replacement of bodies a body that moves out is immediately replaced by another one, so that there is never a void in between the limits, place has been thought to be the limit of a body.

That it is impossible for place to be anything other than one of these, can be established on the basis of a division, as follows.[132] What changes with respect to place, does so either with respect to something in itself or with respect to something external to it. If it changes with respect to something in itself, at all events it changes either with respect to matter or with respect to form. If it changes with respect to something external, it does so either with respect to the extension in between the limits of the container or with respect to these limits themselves. So now that these four suppositions concerning place have been brought to light, if it can be shown that place is neither form, nor matter, nor an extension, then it must be what is left: the limit of the container.

That it is neither matter nor form he has already shown through 548,1
many arguments, and now he reminds us of it. He also adds this, that even if place appears to be form because it coincides with the limit of the container,[133] and because it is neither larger nor smaller, it is nevertheless distinguished from this {form} in virtue of the fact that form is the limit of what is contained, whereas place is the limit of the container. For even if both the form and the place contain while coinciding and being of equal size, nevertheless place is the limit of the air, if that happens to be the container, and form is the limit of

that which is contained. Therefore the place is not the form. But neither is place matter. For even if it receives different bodies while remaining one and the same, and in that respect may appear to be matter, nevertheless these two are again distinguished, because in the case of matter we say that it has become a this here (for what was previously air, that, we say, has now become water), but in the case of place it does not work this way. For we do not say of the vessel – whether of its limit or its extension – that it was previously air but is now water: we say that where there was formerly air, there has now come to be water. Through these arguments and a number of compelling arguments that have preceded, it is surely clear that place can be neither matter nor form.[134]

That it is not an extension in between the limits either, he shows as follows. The meaning of his words is very unclear, and if he had not explained himself in his account of void,[135] his meaning would have remained inexplicable. Different interpreters try to grasp the meaning of what is said in different ways. If, he says, place were the extension in between the limits, then evidently when a body would come to be in it, the extension of the place would pervade the extension of the body.[136] For evidently the place has received the body with its own full extension, so that extensions of the body and of the place have pervaded each other. Now just as the whole extension of the place has pervaded the whole extension of the body, just so the part of place has obviously also pervaded the part of the body. For if the part has not pervaded the part, then the whole has not pervaded the whole. And if the body, and indeed every volume can be infinitely divided, then both the place and the body will be infinitely divided. Hence there will exist an actual infinity.[137] Not only that, but the infinity will be duplicated, for both the place and the body will be infinitely divided. To clarify what has been said let us also set out his
549,1 actual words, which run thus: **If there were some extension which was naturally existent[138] and static,[139] then there would be infinitely many places. For when the water and the air change position, all the parts will do the same in the whole as all of the water does in the vessel**. That is:[140] for when the air gives way to the water and the water comes to be in the place of the air, for example in the vessel, then what has happened to the whole mass of the water in the vessel (and what happened to it is that it has completely pervaded the extension of the place within the vessel), this very thing will also happen to all the parts of the water in all their places, which is equivalent to: these parts too will pervade the parts of the place. As a result these parts too will each be within their own confines and each of them will have its own place.

What follows is a different argument,[141] and it goes like this. If the place of the water is not the hollow surface of the jar, but an extension which is different from the water's extension that is contained by the hollow surface of the jar and that receives the incoming water in the whole of itself, then, when the jar is moved, the jar as a whole will clearly occupy another place which will itself be three-dimensionally extended as well. As a result the extension of place will pervade the whole jar, so that it will also pervade the water inside as well as the extension that has received the water. For the water in the jar will certainly occupy a part of the extension of the place in which the jar has come to be. But it has already occupied a different extension as well: the one in between the hollow surface of the jar. Hence there will be three extensions coinciding: the extension of the place within the jar, the water that has come to be in it, and the part of the place into which the whole jar has moved.[142] The result will be, first of all, that place changes place. For when the jar moves and in its entirety comes to occupy a three-dimensional extension of place, then also the extension inside it which has received the water will be in a place, and if it is thus possible that two or three extensions coincide, then even more of them will coincide, and indeed an infinite number of them.[143] In order to clarify the argument, I will take it up once again and elaborate it with the help of an example.

Let there be a cup,[144] I mean the measure itself, whether of earthenware or of anything else. Now if the extension in between its inner limits is different from the body, for instance of water, that fills it, then when the whole cup is placed in a larger pot, it is clear that also the extension of place in the pot, being three-dimensional, will pervade the whole cup, including its inner volume. It will therefore also pervade the water inside and the extension that has received the water. Hence, three extensions will coincide. Now if I place the pot in its turn in a vessel, the same thing will happen again: the extension in the vessel will pervade all of the extensions of the pot that has been placed inside. Similarly when the vessel in its turn is put into a cask, the extension of the place within the cask will pervade the whole vessel, and so on *ad infinitum*. The result is that many places and many extensions will coincide, which is impossible.

Themistius explains the passage as follows.[145] If, he says, there were an extension naturally capable of existing in its own right and of remaining where it is, there would be infinitely many places. Why so? Because wherever the vessel, filled with a body for instance of water, is transported, the parts of the water will act just as the whole of the water in the vessel. So just as the whole of the water, occupying its own extension, moves together with the vessel when this is transported, just so each of its parts, occupying its own extension, will also move together with the vessel. Now when the jar has arrived at another place, it will obviously occupy the extension in between the limits of the surrounding air, an extension of which both the water in the jar and the parts of the water will occupy a part. Consider then how many extensions coincide.[146] First, that which is occupied by the water in the jar, that is: the extension of the place in between the limits of the jar. Secondly, the extension occupied by the parts of the water; for just as the whole of the water has come to occupy the whole of the extension, so each of the parts of the water has also come to occupy a particular part of the extension and each of them is in a place in actuality. Then again the extension occupied by the jar as a whole; and in addition, the part of the extension occupied by the jar as a whole, in which the part of the jar {which is} the water has come to be. And again, that part of this which is occupied by the parts of the water. If you add to these the extension of the jar itself *qua* body, and the extension of the water *qua* body, and the extensions of their parts *qua* body, you will find many extensions coinciding. And since division of magnitudes proceeds to infinity, the adding of extensions will also go on to infinity.[147]

551,1 For if place is an extension, then necessarily each of its parts too will have to be in its own place in respect of itself.[148] After all each of the parts of the body will occupy a part of its extension of place. For those who claim that place is the limit of that which contains, none of the parts of the body is in a place in respect of itself, not even the parts with which it is in contact with the place, i.e. its own outer limits; but these are all said to be also themselves in place incidentally, in virtue of the fact that the whole is in a place. For none of them is contained by place individually (for they are continuous with the whole) and either they are not in contact with place at all, which goes for the internal parts, or they are not in contact on all sides but only in part; yet place is generally thought to contain that which is in it as a whole. So only the whole is in a place in respect of itself, and through the whole also the parts are said to be in a place. But those who claim that place is an extension necessarily have to say that the parts too are in a place in respect of themselves. For just as the three dimensions of the cubit-sized body pervade the three dimensions of the cubit-sized place, just so the parts {of the body} also pervade the parts {of the place}, and if the parts had not done so, neither would the whole {body} have pervaded the whole {place}. And it makes no difference to me whether someone says that the extension of place pervades the body, or {that} the body {pervades} the place. For it is all the same, to say that one thing pervades the other, or the other way round. Therefore, if each of the parts is in a place in respect of itself, and if the magnitude can be infinitely divided, then there will be an infinite number of places of this same magnitude.

This is what he means in this passage. The next argument Themistius too explains in the same way, namely that if the extension is place, then if the vessel is moved and comes to be in some particular place, it will obviously be in a place that is equal to it. So, if the vessel will be the size of a cubit in all three dimensions, then the place that has received it will also be of that size. Hence, the place that is in between the limits of the jar, which is occupied by the water, will also occupy a part of the place that is occupied by the whole jar; so that a place will be in a place.[149]

And not only will it be possible for two places to coincide, but also, as we have shown, for more than two.[150] For if the extension of place which is occupied by the water, that is: that which is in between the hollow surface of the vessel, is no longer there after the water has pervaded it, then the water in the vessel will no longer be in a place 552,1
either, if indeed this extension is place, and if the extension is not there in the vessel once the water is in it. It will thus turn out that this extension does not exist ever at all. For if someone says that it is no longer there because the water has filled it, then, since there was also some other body in the vessel before the water was there – either air or something else (for it is impossible for it to be void, as will be shown shortly) – it was obviously not there at any other point in time either. Hence it is absolutely necessary that the extension of place in between its inner limits is always there in the jar; and if it is, it is absolutely necessary that when the jar is moved, this extension also moves along and comes to be in a part of the place which the jar as a whole occupies.

These are the attempts culled from the exegetical tradition devoted to the Aristotelian text that are intended to establish that place is not an extension. The external arguments that the commentators have added, and whatever the proponents of the view that place is an extension could say, we will expound after having gone through the text.[151]

211b7 Or some kind of extension in between the extremes.

The extremes of the containing body; for example, if the container is a vessel, the extension in between its inner limits.

211b8 Or the extremes, if there is no extension besides the magnitude of the body that comes to be in it.

If, he says, the extension in between the limits of the container is nothing but the body which comes to be in it, for example water or air, then, since neither matter nor form is place, the remaining possibility is that place is the limit of the container.

211b11 For the extremes of that which contains and of that which is contained coincide.

'Coincide' is equivalent to 'are together and fit onto each other' – which is why place seems to be the form.[152] But even so the distinction is clear: even if both of them are limits and these coincide, still the place is the limit of the containing body, whereas the form is the limit of the one that is contained. Therefore, place cannot be form.

553,1 **211b14** But because that which is contained and distinct is often transferred, for instance water from a vessel.

He states the reason why some have taken place to be the extension in between the limits of the containing body. This is, he says, because just as in a vessel while it remains one and the same several bodies can replace each other, so also, since several bodies come to occupy the same place, the extension appears to be different from the body which comes to be in it. Yet it is not, he says, the extension that is something different; what is different are **the ever different bodies of whatever kind that replace one another**.[153] Therefore there is no extension in between the limits apart from the extension of the body that has come to be in the place.

211b19 If there were some extension that exists naturally and remains static, there would be an infinite number of places coinciding.[154]

He has said **that exists naturally** because of the mathematical {extension}. For the latter does not exist naturally, but in thought. So he says **that exists naturally** as equivalent to 'being and existent', and he says **that remains static**, i.e. immobile, because of the natural bodies, which are seen to be always changing.[155]

211b21 For when water and air change places, all the parts in the whole will do the same thing.

That is: when water and air are replacing each other and yielding to each other in turn their individual places, then what the whole of the water does in the whole of the vessel, all the parts of the water will also do in the parts of the place. So just as the whole of the water pervades the whole of the extension of the place, and the two extensions fit onto each other, so also the same will happen to the parts of the body in relation to the parts of the place. But since magnitudes
554,1 are infinitely divisible, the place and the body will also divide each other into infinity, so that one place will be divided into an infinite number of places; and similarly with the body.

211b23 At the same time place will also be moving; so that place will have some other place, and there will be many places together.

Another argument, viz. that if the extension in between the vessel is different from the body that comes to be in it, then necessarily when the vessel is moved to another place, the extension must move along with it.[156] Hence place would seem to change place, and since the place which has received the vessel has pervaded it completely in all its depth, it has clearly also received this extension inside it. Hence there will be a place of a place, and many extensions will coincide: that which is in between the limits of the vessel, the body that has come to be in it, the part of the extension that has received the whole vessel into which part of the vessel – I mean: the water – has entered.[157] And if it is possible for many extensions to coincide, what will prevent an infinite number of them doing so?

> **211b25** There is no other place of the part in which it moves whenever the whole vessel changes its position, but it is the same; for the air and the water, or the parts of the water, move in the place in which they are, and not in the place in which they come to be.

This is a reply to the second argument, from which it was concluded that place comes to be in a place and that many places are together.[158] For the extension in between the hollow surfaces of the vessel, which received the water, comes to be in another extension when the vessel moves. What he says, then, is that that this absurdity does not follow for those who say that place is the limit of the container. For there is no need, when the whole vessel moves, for a part of it – say, the water – to come to be in another place apart from that place in which it has been all along and with which it moves, and this is the hollow surface of the vessel. The whole has indeed come to be in another place (for it is contained by a different surface of the air), but the part of it – that is the water, or what is in between, or the air – has remained in the place in which it was all along (for it is contained by the same surface, namely that of the vessel), and not in the place in which it has come to be along with the whole. For the water has not come to 555,1
occupy a part of the place in which the whole vessel has come to be; rather, it is in that place together with the whole,[159] but in respect of itself it is not in another place than the one in which it has been all along.[160] By **part**, then, he means a part of the whole vessel filled with either air or water. And a part of that whole is either the air or the water with which the vessel is filled. **For the air and the water, or the parts of the water, move in the place in which they are** – that is: within the hollow surface of the vessel. And **move** is equivalent to 'move along', and **the air and the water** is equivalent to 'either the air or the water', with which the vessel is filled. So the part of the vessel, the water, is in the place where it has been all along, and not in the place in which the whole vessel has come to be. The latter is the part of the air that surrounds it, which place, he says, is a part of that place which is the place of the whole heaven. By

heaven he means the cosmos. So the place containing the vessel is part of the cosmic place.[161] Having said **the air or the water** he added **or the parts of the water**. That is: when the vessel moves, neither the water as a whole, nor its parts, occupy a part of the place of the vessel, but even when the vessel moves both the water as a whole and its parts remain in the very same {place} in which the water as a whole was all along. He has added this to show that the thesis that the parts are in a place in actuality is also something that does not follow for those who say that place is the {containing} surface; whereas according to the first argument, it did in fact follow for those who say that place is an extension. After all, for them it follows that the parts too are in a place in actuality, given that the extensions pervade each other. But for those who say that place is the {containing} limit, it no longer follows. For none of the parts of the whole is enclosed by a part of the {containing} surface, as it would be by a part of the extension. So for those who say that place is the limit of the container none of the absurdities follow that follow for those who say that place is an extension.

> **211b29** Also matter might appear to be place, if one focuses on things that remain at rest and that are not separated.

Having stated whence the idea came that place is form and extension respectively, he now tells us whence the idea came that it is matter. If, he says, someone focuses on a static object in a place, that is: if one
556,1 focuses on the object while it remains static in its place and does not move away from it, and if one then thinks of it as coming to be and perishing while remaining immobile with respect to its place – say, when[162] water which is in a vessel without leaving it becomes air – seen in this way matter would rather seem to be place, since without the object changing place different things come to be in the same spot at different times. For this is how change occurs in matter, without the immanent form changing place. So if one has to compare place and matter, one should think of the water in a vessel as changing into air or into some other body, in such a way that while the whole object remains immobile, different bodies come to be in the vessel at different times. And the words **and not separated but continuous** pleonastically repeat **at rest**.[163] And **continuous** is equivalent to 'attached', for in this case we do not find continuity in the proper sense.

In virtue of the same conception by which matter appeared to be separate from all the forms that come to be in it – namely that the same thing is now white, now black, now this, now that – place too appears to be different from the things that are in place.[164] Therefore, if a thing changes while remaining immobile with respect to place, matter and place may seem to be the same thing. Except that there is a difference: it is matter itself that becomes white and black, or man or horse, whereas place receives these things but does not itself become them. This is also why our common usage differs in these cases: in the case of matter, *what* was water is now air; in the case of place {we say} that *where* there was formerly water, there is now air. And apart from this, as has already been said earlier, matter is inseparable from the object, but place is separable.

> **212a6** By 'the body contained' I mean <the body that is capable of locomotion>.

He says: I do not claim that just every body is contained and in a place, but only that which is such as to be able to be sometimes moving, sometimes not. For this is what he means by **the body that is capable of locomotion**, so as to express: that which partakes of coming to be and perishing. For the heavenly sphere is not capable of motion {in this sense};[165] this is eternally moving, which is why it is not as a whole in a place either.[166] And **locomotion** he uses as equivalent to {motion} 'in straight lines'. For motion in straight lines is locomotion; circular motion is not locomotion, but revolution. Now 557,1
if a thing is capable of locomotion, then it is definitely in a place; but it is not the case that if something is in a place, then it is also capable of locomotion. For the {heavenly} spheres inside that of the fixed stars are in a place, yet they do not exhibit locomotion. In what sense he wants the sphere of the fixed stars to be in a place and in what sense not, we will see him discuss in a moment. To discuss his complete account of these things – I mean the structure of the heavens, of the parts and of the whole – belongs rather to a different work.

(The 'corollary on place' is omitted.)

> **212a7** Place appears to be something profound and hard to grasp, both because matter and form appear to be involved with it, and because of the fact that change of position of the moving body occurs in surroundings that are at rest.[167] 585,5

Having first argued on both sides about place, Aristotle stated that an account of place needs to be offered such that it both solves the puzzles raised with respect to it, and will show that the things that belong to place according to the common conception belong to it in fact, and that it will bring out plainly what is the cause of the difficulty surrounding it. So having given as his account of what place is, that it is the limit of the containing body by which it limits the body it contains, he now wants to show the things he has earlier announced he would show: first of all, that the cause of the difficulty lies in the fact that the theoretical understanding of place implicitly involves matter, form, and extension. The plurality of things thus implied obscures our investigation of place, but extension does so in particular, since it shows a great resemblance to place: it too is deemed to be immobile, just as place in its primary sense is, and to receive different bodies while remaining the same extension. The cause of our mistake is plain to see: extension, as it comes in and goes out together with bodies (through the fact that these – or at least those that fill a place that is the same size – while differing in their other qualities, are the same and equal in extension), is deemed to be one and the same thing and immobile, since its being interchanged escapes notice owing to the instantaneous nature of the mutual replacement of the bodies involved.

What **contributes** to this judgment, Aristotle says, is the fact **that air, too seems to be incorporeal** to most people, due to its fine structure: they imagine that the place in between the extremes is empty, because air does not set our sense-perception in motion in the way that the other elements do.[168]

This, Aristotle says, is why place is deemed to be **not only the limits of the vessel, but also** all **that is in between**: for that is by itself empty. And this is why we need several arguments to show that
586,1 air is a body of sorts, based on the bubbles and noises that are produced when the jar receiving it is filled with water,[169] and on the fact that force is needed to press its mouth under water, and on the fact that water does not flow out of narrow openings as it is held up by the air blocking them, and through the observation that bags filled with air rise to the surface in water and sustain what is placed on top of them, and based on bags being torn and jars broken by the new wine due to the air generated in them, and on countless other things.

So due to the notion that what is in between the limits of the jar is empty, people take it that every place is so too. For place does not appear to be any different from a vessel; the vessel is a **portable place** and place is an **immobile vessel**. The only difference between vessel and physical place then will be in their being mobile and immobile, respectively. This, Aristotle says, is why **when something is moving within something moved**, it is in such a thing {moved} as in a vessel rather than as in a place, for instance as the boat moving in the river: this moves along with the river, so that the water immediately surrounding the boat might be more properly described as a vessel; for place in so far as it is place must be immobile. And, he says, if we must name a place for the boat, the whole of the river might be described as its place rather than that which immediately surrounds the boat.[170] For the river in its entirety is immobile as a whole, so it is rather the whole river that is the boat's place. Place here is obviously not place in the primary sense of the word, but common place, and by the whole river we are to understand, not the entirety of water from the sources to where it reaches the sea, but just the entire bed itself that the water rests on.[171] It follows that also when something is moving in air and the air does not rest in its place but is moved along with it, it may in this respect be said to be as in a vessel, not in a place, but if the air is immobile, then the object is in that as in place: for the place must be immobile while what is in it is moving.

Having thus shown the cause of the difficulty of {grasping} place, he straightforwardly shows on the basis of the account that has been given of it, namely that it is the immobile boundary of what contains, that on that account all the characteristics that apply to place according to the common conception do indeed follow. These were five in number: that it contains and is not anything of the object, that the primary place is neither larger nor smaller than what is contained, that place as a whole has 'above' and 'below', that place is left behind by the individual object and is separable, and that it is immobile. And most of all, that the 'above' and the 'below' belong to place, and also that it be immobile.

In virtue of the fact that place is the limit of what contains, both the centre of the whole cosmos and the limit of the heaven that consists of the hollow surface facing us are places in a most common sense. The one, being the place of the light bodies, appears to be primarily and in the most proper sense 'above' (and this is the hollow surface of the sphere of the moon); the other, being the place of heavy bodies, which is the centre of the universe, is 'down'. By **centre** he means either the centre of the universe, or especially the limit that contains the earth, which is in some regions the surface limit of the air, in others the limit of the water.[172] For the earth is the centre of the universe, so that the centre-place is the one that contains it. What contains it is in some parts the limit of the air verging on the earth, in others the limit of the water that is in contact with the earth.

Furthermore, he says, also immobility belongs to place. For the place 'above' is in its entirety and as a whole immobile both formally and numerically (for even if it moves in respect of its parts, it is nevertheless immobile as a whole according to every type of motion);[173] and the place 'below' is formally always the same, but numerically not any longer.[174] For the parts of the water and of the air are subject to coming-to-be and perishing,[175] and the water and the air are not always in contact with the earth with the same parts. Yet formally it is the same surface that always contains the earth, and as a whole it is again immobile according to every type of motion.[176]

Having said this he says that 'above' and 'below' are said in various ways.[177] For 'above' is both the name used for the place above, he says, i.e. the hollow surface of the sphere of the moon, and the name used for the thing itself above, that is: the body that is contained by {this surface} – he means: everything that is light.[178] In its turn, 'below' is both the name used for the surface of the air verging on the middle of the universe, i.e. the surface that contains the earth, and the name used for the bodies themselves in the middle, i.e. those that are heavy.

The following as well was a characteristic of place: that it is neither larger than that which is in place nor smaller, and this as well is part of this account of place. For if it is the limit of what contains, and the limits of the container and of what is contained are together (for the surfaces fit onto each other), then the place is neither larger nor smaller, but equal in size. Equal in size of course with respect to the circumference – for it is not equal with respect to the whole three-dimensional extension. In consequence those who say that place is an extension may have a more reasonable way of saving the idea that place is equal in size to the thing that is in it. For on this view it will be equal in every dimension.[179] Even more clearly can the claim that place is immobile be saved on this view.[180] The limit of the container moves along with the body of which it is the limit, but the {three-dimensional} extension, being void in its own right and not a substance, is unchangeable according to every type of change. Hence, on the former view no place is immobile, but on the latter view every place is. For place does not simply need to be formally immobile, but also numerically. And perhaps on the former view it cannot even be formally immobile. For if a thing is immersed in a river, and then when the water of the river sinks it is no longer the surfaces of the water that contain it but those of the air, how come that the place has not been swapped and changed formally as well, though the object {in it} remains immobile?[181] They reply to this in defence that one should not consider place to be the limit of the container *tout court*, but the limit together with this particular relation towards the thing contained.[182] So just as things growing, and in general things that are fed, remain numerically the same in virtue of the fact that the same form is preserved even though their matter is in flux, just so place as well will remain one and the same because what is contained will be kept in the same relation {to it}. But this is surely rather absurd. For in this way the result will be that a thing can never come to be in ever-different places. For when the air is immobile, and we are moving, then necessarily, in whichever part of it we come to be, we will have the same relation towards it as the one according to which we are precisely contained by it. So Socrates will be in one and the same place when he is in the market place and when he is in the Lyceum. Hence, he has not changed place. Therefore, if place is the limit of the container, it is impossible for place to be immobile, but it has to be moving even if the object {in place} is at rest. As a result, it is necessary also for this reason that place is extension.[183]

And if someone says: 'but it is not moving *qua* place, for it is a surface, and that is in its own right immobile, since it is also incorporeal', I reply to this that the surface is not place without any further qualification, for then every surface would be a place, but only in so far as it contains; and as something that contains it changes.[184] For if, while the object that is contained remains immobile, {the surface} at one moment contains and at another does not, or if it now contains this, now that, then it is clear that {the surface} itself is what is moving, moved along with the body of which it is a limit.

So much for these things. Aristotle has not added the remaining two of the characteristics belonging to place – that it contains the object without being anything of it, and that it is left behind and separable – since they are obvious and plain to see. For if place is the limit of the container, then from this very definition it is clear that it contains, that it is separable, and that it is nothing of the object. After all what is in a place is separated from this place, but the limit of the container is not separated from that whose limit it is, and yet the 589,1 limit of the container is not anything of what is contained either – neither a part nor an affection nor anything else. These very same characteristics, except that of containing, can also be saved by those who say that place is an extension. For the extension too is different from the body that is in it, and is separated from it, but that it should contain is something they will either not admit at all as a characteristic of place, or, more likely, for them too containing will apply to place in the same way in which it applies for the other party. For those who say that place is the limit of the container think that it is clear that place contains, because the limit belongs to a particular body, and because the limit contains what is in place from the outside – but they think so focusing not on the limit, which is place properly speaking, but on the body of which it is the limit. But just as the limit, which is place properly speaking, while fitting onto the limit of what is in place, and while being neither larger nor smaller than it but coinciding with it, is nevertheless said to contain the thing, just so it is clear that the extension too, which with its own limit fits onto the limit of the body that is in it, may in this respect also be said to contain this body. For by what arbitrary choice, if the limits of place and of what is in place fit onto each other on the one view just as well as on the other, would we say of the one that it contains but not of the other? That {place} contains should therefore either be maintained on neither view, or on both.[185]

212a9-10 And because of the fact that the replacement of the moving body occurs in a container that is at rest.

For while the vessel remains at rest, that which is contained in it is being replaced, and ever-different things occupy the same place. Yet not only in the vessel, but also where the surrounding air remains immobile, ever-different bodies come to occupy it. Since, then, the container while remaining at rest receives ever-different bodies, the extension within the container appears to be an extension in its own right, different from the extensions of the bodies that come to be in it.

> **212a14** And just as the vessel is a portable place, so the place is an immobile vessel.

Since he has elaborated his account focusing on the vessel – and in order to avoid that someone might say 'and what bearing does it have on place if things are this way in the case of a vessel?' – for this reason he says that place and the vessel differ in no other respect than by being mobile and immobile respectively. This is why he even uses the two terms interchangeably and both calls place a vessel and the vessel a place, so that people may say about place as well the things one might say about a vessel, and, conversely, about a vessel what may be said about place. This is why he says that even when a thing is not in a vessel but in a place, when that in which it is located is moving – such as the river with the boat, or the air with a wing, and especially if they move with equal speed – this thing **uses** its container **as a vessel**, not as a place.[186]

> **212a18** Place is meant to be immobile.[187]

There you are: just as I said at the beginning, he here adds the remaining one of the presuppositions about place, viz. that place should be immobile and not move along with what is in it.

> **212a20** So: the first immobile limit of the container, that is what place is.

From what has been said he concludes that what place is, is the primary limit of the container, being immobile. **Primary** is equivalent to 'immediately fitting that which is contained and {precisely} in so far as it contains'. Hence, even if the vessel is said to be a portable place, it is not a place in so far as it is portable, but in so far as it is a {containing} body. For not all the vessel is a place, but its inner limit is, i.e. its hollow surface, which is immobile in respect of itself, but moves incidentally in virtue of the fact that the body whose limit it is, is moving. For if the parts {of a body} move not in respect of themselves but incidentally when the whole is moving, how much more does this apply to its limits. So every place *qua* place is immobile in respect of itself.[188]

> **212a21** And it is for this reason that the centre of the world and the extreme limit, on our side, of the revolution appear to be the one 'above' and the other 'below'.

From here he goes on to tell us that the characteristics that apply to place according to the common conception can also be shown indeed to apply on the basis of the account that has been given of it. Because place, he says, is the limit of the container, and there is both 'above' and 'below' in place, for this reason all people think the hollow surface of the sphere of the moon is 'above' (for that is where light bodies 591,1 move to), and the centre of the whole is 'below' (for that is where heavy bodies move to).[189]

212a23 Because the one is always at rest

That is: the centre of the world, which is the place of the heavy bodies. This **is always at rest**, i.e. is immobile, and it changes numerically but not formally.[190]

212a24 And the limit of the revolution remains in the same state.

That is: the hollow surface of the sphere of the moon. For this, even if it does change with respect to its parts, is yet in its entirety and as a whole immobile, being one and the same both formally and numerically, and always in the same state.[191]

212a24 So that since the light is that which by nature moves upward.

Since there is that which by nature moves upward (this is what is light) and that which by nature moves downward (this is what is heavy),[192] 'above' and 'below' can both be two different things. He has added in what sense they are two different things. **The containing limit facing the centre**, he says, **is 'below' and so is the centre itself**, using **centre** to refer to the earth, and using **the containing limit facing the centre**' (that is: what contains the centre which is the earth) to refer partly to the surface of the air where it is in contact with the earth, and partly to the surface of the water. Similarly, also the 'above' is two different things: the extreme part of the heavens that faces us (that is: the hollow surface of the sphere of the moon) and also the extreme part itself, that is: the light body, i.e. fire.[193] This is how some people have explained these passages, as we have already said before in the theoretical introduction to the lecture.[194] Themistius, however, explains it as follows.[195] Upward and downward motion, he says, is something that happens to bodies by nature: for light bodies movement is by nature upward, for heavy bodies by nature downward. The place 'below' is constituted, he says, by a set of three bodies of sorts – the centre itself with those {elementary bodies} that surround it – and these are earth, water and air. For each 592,1 of the heavy bodies is contained by these elements, either by all of them or by some of them. They are contained by all of them, he says, as in the case of rocks sticking out of rivers (for these are contained partly by the earth which is at the bottom of the water, partly by air, and partly by water). People walking or swimming are surrounded by two of them: the one set – those who are walking – by earth and air, the other – the swimmers – by water and air. There are also things that are contained by {just} one of these elements: birds in the sky by air, fish swimming under water by water, and by earth certain parts of the earth that are distinct from it yet in contact with it and are surrounded by it on all sides. This explanation appears to be more suitable.

Some have taken **the centre of the world** to be {literally} the centre of the universe,[196] but these people for sure go about in an offhand fashion. For the centre is not the place of anything at all – on the contrary, all things move as if to the centre, yet the centre is {itself} not the place of anything, whether place is the limit of the container or the extension in between the extremes. With regard to the former alternative it should be noted that, on the basis of these considerations as well, it is shown that place is not the limit of the container. For if place is the limit of the container, the same things will be the places of each other.[197] For the place of the earth is the surface of the water that is in contact with it, and also that of the air (for these constitute the place of the earth, namely the air and the water, containing it from all sides, one this part, the other that), or the limit of air and water on the side of the earth. But since water as well is among the bodies that are in a place, there will be a place of it, and this not in a singular way. For it is not only the air that is in contact with the water from all sides, but in some parts this is air, in others it is earth. And so in that part of the earth where the place of the earth is water, or rather the latter's surface, there the earth as well will be the place of the water. Hence the earth and the water will be each other's places in virtue of the same thing. So it will result that the same object will in virtue of the same thing be both place and in place. If these things are impossible, place is not the limit of the container, but rather the extension that encompasses each physical body, which is void in its own right, yet never actually remains empty but is filled by the body that comes to be in it and whose place it is.[198]

593,1 **212a28** And because of this, place appears to be a surface, and like a vessel, and a container.

Not because of what has been said immediately before this, but because it is the limit of the container. For if it is a limit, place will obviously be a surface. And if place contains from all sides, and a vessel which contains what is in it from all sides is similar in this, then place resembles a vessel too.

212a29 Moreover, it is somehow together with the object, for the limits are together.

That is: it is neither falling short of it, nor too large. For if it is the limit of the container, it evidently fits onto the limit of what is contained, and, fitting onto this, it is together with it and coinciding, neither falling short of it, nor being larger.[199]

Chapter 5

212a31 If a body has another body outside it and containing it, it is in a place. If not, {it is} not.[200]

Having said which things are in a place, Aristotle goes on to say which things are not in a place.[201] For if, he says, place is the limit of what contains, then if a thing does not have a body outside itself, it is not in a place; for it will have nothing outside it that contains it. Hence the universe is not in a place: there is nothing outside it. For suppose it has something outside it, then it can no longer be in a proper sense the universe.[202] This is why the heavens, if there is nothing outside them, as indeed there is not, cannot be in a place either. So just as, he says, even if the universe were a single substance, for example water (which admittedly is {normally} in a place; for it runs away and is by nature not able to keep together but in need of something that keeps it together and confines it), then although its parts (that is, if they were divided in just the way the parts of the universe are divided now) would themselves be in a place, yet the whole of the water would not be in a place since it would not have a body outside it that contained it – just so, he says, now too, the universe, being the way it is, is not in a place, since there is nothing outside it.

And if the heavens are not in a place, how do they change with 594,1
respect to place?[203] Well, he says that they are said in a way to move, in a way not. They do not move in respect of themselves as a whole, for they do not exchange their place as a whole. And this is what change with respect to place of whole entities amounts to: when the thing as a whole exchanges the place in which it is, for example the things that move in straight lines. So the heavens, just as they are not in a place, do not move in respect of themselves as a whole, but they do rotate, and a motion of this kind is a motion of the parts.[204] For in the process of rotation, the whole remains immobile as a whole, yet the parts, which yield their places to each other, it is these that exhibit change with respect to place, and the whole is said to move with its parts.[205] For when all the parts move, the whole as well is said to move, not in the primary sense, but as I said, in respect of its parts. According to Aristotle, it is the parts themselves that mutually provide each other's place, in virtue of the fact that each of them is

contained by those immediately surrounding it (for the parts are like places to each other in so far as each of them is surrounded on all sides by its neighbours).[206] According to his exegetes, however, the place of the sphere of the fixed stars in respect of its parts would be the convex surface of the sphere inside it, and different parts of the one sphere are in contact with different parts of the other sphere at different times.[207] It is clear that we call this a place by analogy, for it does not contain from the outside, yet it does exhibit some resemblance to place in so far as it marks off. For just as place properly speaking marks off an individual thing from all other things, just so the convex surface of the inner sphere as well marks off the body on its outer side from all other things. In this respect, then, the convex surface of the inner sphere may by analogy be the place of the sphere of the fixed stars, and the place of the whole of that sphere in respect of its parts: for it is these that cover ever different parts of that surface. And what has been said about the sphere of the fixed stars can also be said about the universe {as a whole}: for the universe is itself not in a place (for there is nothing outside it), but in respect of all of its parts it is in a place. For some of them – the bodies that are eternal – rotate; others, the bodies that are subject to generation and perishing, move upwards and downwards.

Having said this, Aristotle wants to set out more clearly in what way the heavens are not in a place and in what way they are, and makes a division of things that are in a place. Of the things that are in a place, he says, some are in a place potentially, others in actuality. All things that have their own boundary marked off by their place from all other things, are said to be in a place in actuality. The parts of a whole that is continuous and in a place are said to be in a place potentially. Potentially, because they are capable of being separated
595,1 from the whole and being in a place in their own right, and of being no longer continuous, but in contact, like a heap of wheat. For each of the grains, while being a part of the whole, is in a place, because they are not continuous, but each is within its own boundary.

Furthermore, of the things that are in a place some are in a place in respect of themselves, others incidentally. In a place in respect of themselves are all those things that are capable of locomotion or of increase or decrease, and these are the bodies that are subject to generation.[208] All the accidents of things are said to move incidentally, and so is the soul, whether {taken as} in the body as in a substrate or as a separate entity.[209] For these things are never in a place in respect of themselves, for they are incorporeal. Given that things are thus said to be in a place in four different ways, he says the heavens are to be ranked among the things that are in a place incidentally. For they are not in a place either in actuality (for they are not contained by anything from outside) or potentially (for they will never be contained; and things that are in a place potentially are such as are able to come to be in a place in actuality at some time). But neither are the heavens in a place in respect of themselves (for they are incapable of locomotion; for what moves changes its place as a whole, but the heavens remain in the same position and move in respect of their parts, and the motion involved is a motion of their parts, not of the whole). The only remaining possibility is that the heavens are in a place incidentally.

With respect to this solution, Themistius raises the puzzle how to avoid Aristotle's contradicting himself, since he now calls motion in respect of the parts incidental motion, whereas earlier, in the passages in which he was trying to establish that it is possible for something to be in itself as in a place, he declared that a thing can be in itself in respect of its parts but cannot be so incidentally, presumably on the understanding that 'in respect of its parts' is not the same thing as 'incidentally'.[210] He solves the puzzle by saying that Aristotle probably used 'incidentally' here in the more general sense as equivalent to 'in respect of something else' (for it is his habit often to put species in place of genera), since 'in respect of something else' is a wider term than 'incidentally' that can be used in the case of what is in respect of its parts as well as of what is incidentally. For it can be used both in the case of the 'incidental' and in the case of what is 'in respect of something else'.[211] Or perhaps it is safer to say that he used the word in a non-proper sense. For just as we often say, when a whole is moving, that the part is moving incidentally, and when the whole is in a place, that the part is in a place incidentally – even if one cannot properly say, if the whole {body} is moving, that the hand is moving incidentally, but {ought rather to say} that it is moving in respect of itself, yet neither separately nor in so far as it is a hand, or more unqualifiedly a part, but {only} in so far as it is unified with the whole {body} (which is why it could more properly be said to move not in the primary sense but in respect of something else, for the motion of the parts takes place in respect of the whole) – just as, as I said, when the whole is moving we often improperly say that the part is moving incidentally, just so, when the parts are moving, the whole could also be said, in a non-proper way, to be moving incidentally. For the whole might more properly be said to be moving in respect of something else when its parts are moving, because what is moving in the primary sense are the parts.

In this way the heavens as a whole may also be said to be in a place, not incidentally (for they are not an accident of something else that is in a place in respect of itself; and it is things of this kind that are wont to be {something} incidentally), but rather in respect of something else, in virtue of the fact that its parts are all in a place. But since Aristotle has said that some things are in a place in respect of themselves, it is reasonable that he opposes 'incidentally' to 'in respect of itself', yet includes in this use of 'incidentally' also what is

15 in a place in respect of something else. For having said **others incidentally**, he has added **like the soul or the heavens**, one of which is in a place incidentally (the soul) but the other in respect of something else (the heavens).

Having mentioned the things that are in a place potentially and those that are so in actuality, and the things that are {in a place} in respect of themselves and those that are so incidentally, he should also have mentioned the things that are in a place in the primary sense and those that are so in respect of something else.[212] However, the things that are in a place in the primary sense he has {here}
20 included both in the things that are in a place in actuality and in the things that are so in respect of themselves; and the things that are {in place} in respect of something else {he has included} both in the things that are so potentially and in those that are so incidentally. After all, the things that are in a place in actuality – I mean the things that are not continuous with other things but that have their own boundaries marked off – are in a place in the primary sense, and so are the things that are {in place} in respect of themselves, such as the things capable of locomotion, as he himself has said, i.e. the things that
25 exchange their places as wholes. It is these things that are in place in the primary sense; and so their opposites – what is in place either potentially or incidentally – would include also what is in place in respect of something else. For what is in a place potentially, for example the parts, is in place in respect of something else and also in respect of itself (for in virtue of the fact that the whole is in place in respect of itself,
30 the part is surely in place in respect of itself as well).[213] Yet the things that are in a place incidentally, are not necessarily also in a place in respect of something else; examples are the soul, and whiteness. For these are in a place incidentally only. This, then, is why he has through his examples included the things that are {in a place} in respect of something else along with the things that are {in place} incidentally.

So the heavens as a whole, he says, are in a place in respect of all their parts (for each of the parts is contained by the parts surround-
35 ing it, and the parts are like a place for each other); yet the universe
597,1 is not in a place. For if it were in a place, it would have what contains it from outside. In this way it would not be the universe or the whole, if it had something containing it. **And for this reason**, he says, we say that **all things are in the heavens,** because the heavens are the universe:[214] for they contain everything from outside, and have included it within themselves.

5 Having said how the cause of the unclarity has become plain from the account that has been provided, and having shown that the characteristics that apply to place according to the common conception show themselves as present on his definition too, Aristotle goes on to show that, in addition, all the puzzles that are raised with regard to place are solved on the basis of the definition of place that has been provided.

The first puzzle claimed that if place exists, then if body is three-dimensionally extended, place must be three-dimensionally extended as well. But this is {in fact} how all {that is} body is defined. Hence body will pervade body. Now if it has been shown that place is not an extension, but the limit of what contains, the puzzle is solved, or rather, it will not even arise in the first place.

The next puzzle claimed that if there is a place of a body, there will also be a place of a surface and of a point. But it is not necessary, he says, that if a body is in a place, then also a surface and a point will be in place. For these are neither separable, nor in a place in respect of themselves. For if we say that not even the parts {of a thing} are in a place in respect of themselves, but that it is in virtue of the fact that the whole is in a place that these too are said to be in a place, on account of the whole, how much more should we say the same thing about surface, line and point? For the limit of what contains the whole thing as a whole is its place, not {the limit of what contains each} of the parts individually.

Another puzzle claimed that, if place exists, then necessarily if bodies grow, place will have to grow along with them. If place were an extension, he says, it would necessarily in its own right grow along with the body that occupies it. But in fact, as being the limit of what contains, it will not grow along with what grows. For the limit will not grow in its own right (for growth is a feature of bodies). Rather, when in the process of the growth of what is in place the surrounding body will make way for it, whether through mutual replacement or through compression and condensation, then its limits also will recede, and it is in this way that bodies that grow come to be in a larger place. Hence, the place has become larger, but it has not grown in any way. Also, since bodies do not grow from what is in no respect and in no way anything, but through assimilating food, in the process of which the food changes into the parts of the body, it is clear that the body that has grown takes up just that place which was occupied by the food. Hence it is not necessary either that there is a void or that there is some body which is not in a place.[215]

It is plain, however, that those also who claim that place is three-dimensionally extended will say these same things. And much more rightly so, in so far as, when the growing body increases in size and the surrounding body is extended as a result, the surface of the body becomes larger than it was before, even if this does not occur through growth, but, as I have said, through the body of which it is a surface being extended. It is for this reason that one might suppose it to be growing; on the other view not even the barest suggestion is left that place grows. For since the extension is separable from all body and for that reason immobile, when the body has grown and has pushed away the body next to it, whether this be air or water, ever so little, it has come to occupy the place of the latter, while the extension has remained completely unaffected; and thus the thing that has grown has come to be in a larger place, without place {itself} having undergone any change at all. And if what is added to the growing body comes to occupy the place of the food, what necessity is there for the {underlying} extension to grow? It would only be necessary for place to grow along {with the growing body} if place were the form of what is emplaced, for this is what grows, as is shown in *On Generation*.[216]

But neither is it necessary {on this view} that there is a place of a surface, nor of a mark, or point. For if what is in a place in respect of itself is the body, what necessity is there for the limits of this body to be in place {as well}? For if we have shown that not even its parts are necessarily in a place in respect of themselves, how much more does this hold for its limits?[217] And if it is in virtue of the fact that limits fit onto each other that people think they must be in place as well, then why will this follow any more on the present hypothesis than on Aristotle's?[218] For there as well limits necessarily fit onto each other. And if a vessel does not differ from a place in any other respect than by moving, and the vessel is filled by the air, or the water, that is in it, then the place too, while being void on its own account, is filled by what is in place. But a surface does not fill a surface, nor does a line a line, nor a point a point, and they are not void on their own account either; they merely fit onto each other.[219] So it is clear, then, that there is no need at all to claim that there is a place of surface, or line, or point.

599,1 So much for these things. But Aristotle also solves Zeno's puzzle. 'If everything that exists is somewhere', the latter said, 'and if place exists as well, then place too will be somewhere. Hence there will be a place in a place, and so on *ad infinitum*'. This one is solved as well. For even if it is necessary, he says, for everything that exists to be somewhere, and if for that reason place too will be somewhere, then let place be somewhere, **but not as in a place**, but as the limit in what is limited. **For not everything that exists is in a place**, he says, **but only the body that is capable of locomotion**,[220] i.e., that is by nature capable of both motion and rest; in other words, what moves in straight lines. For everything of this kind is necessarily in a place, but not everything that {moves} in a revolution is in a place, but only the inner spheres {of the heavens} – and now he does in fact take the system of the eight spheres as one continuous whole.[221]

The following, too, **reasonably** applies to things that are in a place: that each tends to move to its proper place and remains in it. And this he establishes very elegantly.[222] He makes the following presupposition: everything, he says, that is in contact with something by nature, **not by force**, is **akin to** what it is in contact with. He has plausibly added 'not by force'; for a clod of earth that is suspended is also in contact with the air, yet not by nature but by force. As has been said in *On Generation*, those things are in contact with each other that have a common tally linking them to each other.[223] This is why earth is in contact with water (for they have the cold as common tally), water with air (here what is common is the moist), air with fire (what is common is the hot), and fire with the sphere of the moon (there the community is only by analogy: both are fast-moving and alight).[224] And things that are in contact are akin because they change into each other easily, as has been said in *On Generation*.[225] Now since each thing, when it has been separated from its own wholeness, strives to regain this (for everything desires what is akin {to it} as well as its own wholeness) and for this very reason also {its own} being; and each thing *is* first and foremost when it is in its own proper wholeness, <or> when it is moving towards its own wholeness,[226] for its motion and its striving have this as their preferred object, and towards its own place. For this is akin to the whole. But, someone will say, 'even if what is in contact is by nature akin, and the air is in contact with water on its lower side, its parts must move downwards.' Now that air is akin to water in so far as it has the moist as a common tally, is clear. But we advisedly said that each thing prefers to move towards its own wholeness: the wholeness of air belongs with the things that are upward-moving; air is more akin to fire than to water, yet it spreads in all directions, not because it is by nature such as to spread in all directions, but owing to the force of the void, due to which even water often becomes upward-moving and rides upon air, as in the case of clepsydrae.[227] Hence even when air is in contact with water, it still is so due to the force of the void, since by nature it is upward-moving. Each thing, he says, also reasonably remains by nature in its own place, since the part remains in its own wholeness too. **And what is in a place**, he says, is as **a part is in relation to a whole**. For what is in a place has the same relation to the place as a distinct part has to a whole. Just as if one dips one's hand into water and divides a part of the water {from the remainder}, this is also the relation that what is in a place has to what contains it. For the container is like a whole, and the contained like a part. After all, he says, the container stands to the contained as a compound body stands to its matter. Water, after all, is the matter of air, and air is some kind of form and actuality of water, and as it were a compound body. This is also why the change from water to air is a species of generation, as has been said in *On Generation*, but the change from air to water a species of perishing.[228] And both the form and the matter are deemed to be parts of a compound. Hence if water is analogous to matter, and air to the form or to the compound, and matter is part of the compound, then it turns out to be correct to say that what is contained stands to the container as a distinct part does to a whole. And a part tends to remain within its own wholeness; it is reasonable, then, that each thing also remains in its proper place, that is: {within} its container, which stands in the same relation to what is contained in it as the whole does to a part.

212a35 It rotates; for that is the place of its parts.

People try to find out what kind of parts Aristotle refers to here: are these the parts of the whole cosmos, which are merely in contact with each other as discrete entities, or are what he calls parts the continuous parts of the outermost sphere?

601,1 Some say that what he calls parts are the spheres that are in contact with each other, and the bodies that are next.[229] What he says, then, is that the whole is not in a place, but the parts are the things that are in place. Now in order to indicate what is the place of the parts, he has added **it rotates; for that is the place of the parts** – meaning, not that it is in virtue of the fact that the outer sphere is rotating that it is the place of the parts next {to it}, but with the purpose of indicating that it is not continuous with the bodies that are next to it, but discrete. For this is also why it rotates on its own account. Now in virtue of the fact that it rotates – in other words, in virtue of the fact that it is divided from the things that are inside it and that it contains them – in that respect it is also the place for its parts. That by 'parts' of the universe he refers to the bodies that are contained on the inside, is clear from what he has added: **some do not move upwards and downwards, but rotate; some – those things that exhibit condensation and rarefaction – do move upwards and downwards**. If some of the parts of the whole move upwards and downwards – and these are the things that are subject to generation {and perishing} – and others rotate, it is reasonable that what he calls the 'all' is the whole cosmos, and that what he calls its parts are the elements of the cosmos, i.e. the spheres and the bodies that are within these.

Others say the parts {Aristotle is referring to} are the continuous parts of the outermost sphere. For that sphere is neither as a whole in a place (for there is nothing that contains it), nor are its parts individually in a place in respect of themselves (for they are continuous with the whole); and neither does the whole sphere move as a whole (for it does not change its whole place), nor indeed do its parts move *qua* parts (for they are continuous),[230] but it moves with its parts, and the place for these parts, as we have said, is either the containing parts surrounding them, or, more properly, the convex surface inside it of the sphere of Saturn – by analogy, as we have said.[231]

By the same analogy they equally speak of this surface as the place of the fixed sphere as a whole. For it marks this off everywhere from the inside. To support this exegesis one should take the words **it**

rotates, for that is the place of the parts as follows. Having said that **it does not change its place as a whole at once**, whereas he should have said 'but it moves in respect of its parts', he actually says
it rotates. But this is the same thing; for rotation is motion in respect 602,1
of parts. And the words **for that is the place of the parts** are equivalent to 'the circle {is the place of the parts}'. For the parts contain each other in a circle.[232] Alternatively, since it moves in rotation around the convex surface of the sphere of Saturn, he meant that this place, the convex surface of the sphere inside it, is the place of its parts.[233]

> **212b7** And some things {are in place} in respect of themselves; for example every body that is capable of change in the form of locomotion or increase is somewhere in respect of itself.

He says that the things that move in a straight line, the ones that he says are also in a place in respect of themselves, exhibit locomotion; the heavenly spheres, on the other hand, do not exhibit locomotion but they rotate. For locomotion differs from rotation. The one is motion {of a thing} as a whole, the other in respect of a part. He takes the system of the eight spheres as a single thing, which is why he says that the heavens as a whole are not in a place.

> **212b10** In so far as it moves, in that respect there is a place for its parts.[234]

He has said this both earlier, 'and it rotates; for that is the place of its parts', and here he says the same thing: 'in so far as it moves' – i.e., in so far as it does so in a circle – 'in that respect there is a place for its parts'. For in virtue of the fact that each of the parts of the heavens, i.e. of the system of the eight spheres, rotates, in that respect each outer part is the place of the next part inside it. So by 'parts' he is likely to mean each of the spheres. What follows also makes this clear: **for each part**, he says, **is consecutive to another one**; and what is **consecutive to** something else is, as he himself puts it in book five, 'what is successive to something else and in contact with it'.[235] And this is also how the spheres are related to each other. Still, the outermost sphere is not in a place, unless it is, as has been said, in an analogical way, in virtue of the convex surface of the sphere inside it.

> **212b11** Other things {move} incidentally, like the soul and the 603,1
> heavens.

This phrase links up with what was said {a few lines} earlier. For having said there that 'some things are {in a place} in respect of themselves, for example every body that is capable of change in the form of locomotion or increase', he has here stated the counterpart by saying **others** {are in a place} **incidentally**. The other things that are said in between are about the heavens. For after having said that the things that move in straight lines are in a place in respect of themselves, he has added next how the situation is for the things that rotate, and then he has here provided the counterpart of this account. He says that the soul is in a place incidentally, and the heavens too, as a whole, of course. For in virtue of the fact that the parts are in a place, and that the whole is in its parts, the whole may also be said to be in a place incidentally, or rather, in respect of something else.[236]

212b12 For the parts are all in a place in a sense.

He did well to add 'in a sense'. For not all the parts are in a place in the proper sense. For the outermost sphere is not in a place in the proper sense, and neither are parts that are continuous.[237]

212b13 This is why only its upper part rotates.[238]

Since the heavens are not in a place, otherwise than incidentally, as has been said, or rather in respect of their parts, they also only exhibit rotation, but in no way motion in a straight line. For things that move in a straight line are necessarily in a place. So the heavens, not being in a place, reasonably only exhibit rotation.

212b14 And the universe is not anywhere.

Having said that the heavens are in place in respect of their parts, he now shows that it is impossible for them to be in a place as a whole in respect of themselves. For what is in place, he says, is itself something, and has outside itself what contains it, which indeed is its place. Now if the heavens are a whole and are the universe,[239] there can be nothing that serves as a container of them from outside, since it would not then be all {there is, i.e. the universe}'.[240] For outside the universe there is nothing.

604,1 **212b17** And this is why all things are in the heavens. For the heavens possibly are the universe.

The time-honoured opinion about them, he says, that all things are in the heavens, is also in accordance with our theory that the heavens are not in a place. For if all things are in the heavens, then nothing is outside them, and if there is nothing outside them, they cannot be in a place. For a place needs to contain from outside what is emplaced. And having said that **the heavens are the universe**, he has added **possibly**, because he has not yet proved that there is nothing outside the heavens.[241]

212b18 And place is not the heavens, but an inner immobile limit of the heavens that is in contact with the body that is capable of locomotion.[242]

Since he said that 'all things are in the heavens', because in his view the heavens are the most general place of all things, he has added in what sense the heavens are a place, i.e. that they are not so in respect of themselves as a whole, but in respect of their inner limit, with which they are in contact with the body that is capable of locomotion, that is: {the body} that has the potentiality to move and not to move; i.e.: that which is capable of generation and perishing. In what sense he calls the limit of the heavens that faces us, which is indeed place, **immobile**, has been said in what precedes.[243]

212b20 And this is why the earth is in the water.

He explains in what way all things are in the heavens, viz. not in virtue of containing all things immediately, but in virtue of the fact that they are the ultimate container.

212b22 It is clear from all this that the puzzles also are all solved.

From here onwards he solves the puzzles; we have discussed these sufficiently.[244]

212b27 But not the extension of a body.

That is: {not an extension} receptive of a body; or rather: not {one that is} body-like, i.e. three-dimensionally extended, for a surface and a line are also extensions.

212b31 And if they are fused, they are not capable of being 605,1
acted upon, but if they are in contact, they can act on and be acted upon by <each other>.

Since parts too appear to be successive and in contact with each other, he sets out the difference between what is as in a place and what is as in a whole. For they share the property of being next to and in contact with each other, yet when these things are **fused**, then they are **not capable of being acted upon** by each other, and are as a part in a whole; but when they are merely in contact, and not fused, then the things contained are in the things containing them as in a place, and the things containing and contained can act on, and be acted upon by, each other. Things that are as in a place differ from things that are as in a whole, in virtue of the fact that things that are as in a whole are continuous with these wholes and {the two} are incapable of being acted upon by each other, whereas contrariwise things that are as in a place are divided from what contains them and are merely in contact with it, and {the two} do act on, and are acted upon by, each other.

> **212b33** And what is in a place is like a separable part in relation to the whole.

Since he has said that it stands to reason that each thing remains in its own place, because a part that has come to be in its own wholeness remains there too, he wants to show that things that are contained are also like parts of what contains them. And by parts he does not mean continuous parts, but {parts} that are actually divided. As a clear example he gives what happens to water.[245]

> **213a1** This is how air as well stands to water; for it is matter, and the other is form.

Just as a part of water that is divided from and in contact with {the body of water it is divided from} relates {to the whole}, just so water stands to air, and air to the fiery region. And in what way it is a part
606,1 he has shown next by saying that water is **matter** to air, and that air is a kind of form supervening on water. Air, then, is something like a compound body, compounded of matter and form, and the matter too is like a part of this compound.

> **213a3** And air is potentially water in another way.

For even if each of them is potentially what the other is, still, he says, it is not so in the same way. For it has been said in *On Generation* that the change from water to air is generation, and the reverse perishing. So air is potentially water, but in the sense of being brought towards perishing and not-being {as it becomes water}; and water is potentially air, being brought as matter to a form. In *On Generation*, he says, we have made the relevant distinctions concerning these matters in greater detail; now, he says, we have only mentioned them in so far as the need arose. And this need did arise because he wanted to show that things in a place stand to place as a part does to a whole: in order to show this, he said that water stands to air as matter to form. The theoretical account of this, he says, we will set out in greater detail elsewhere.

213a6 If, then, matter and actuality are the same thing (for water is both, but the one potentially, the other in actuality)

If, he says, water is both matter and form – potentially form, but matter in actuality – and air is clearly the form of water, then, when it has come to be air, it is a compound thing. Hence, since matter is a part of the compound, water will also be a part of the air. And he did well to say **in a way**: for it is not a part in the proper sense, but analogically. For this reason, he says, since it stands to air as a part, it is by nature in contact with it, but is not fused with it. But when it changes into air, then it gets fused with it and becomes a part of it in the proper sense.

Notes

1. This first section of the commentary – no doubt corresponding to a first lecture, on which see the Introduction, pp. 1-2 – runs from 496,1 to 504,8 in Vitelli's edition. It covers Aristotle *Phys.* 4, 208a34-b27, where Aristotle, after a brief introduction on the importance of the subject, starts his discussion by setting out a number of possible reasons to assume that place exists. This is the first part of his dialectical investigation – setting out the *phainomena*, or the 'things as they generally appear to be the case' – and the rules of his dialectical method do not require Aristotle to accept all of these *phainomena* as true or even plausible (on which see Algra (1995) 170-81). As the main points that emerge from Aristotle's survey, Philoponus seems to mark the tenets that place seems not only to exist, but also (1) to be different from the emplaced body (498,28-30), (2) to exhibit differentiations ('up' and 'down') and a certain power (499,1-2), and (3) to be somehow prior to everything else (501,12).

In the first parts of this commentary Philoponus appears to have primarily set himself the task of elucidating Aristotle's text to the best of his abilities. His critique appears to have been primarily reserved for the *Corollary on Place*, which is inserted into the commentary on *Phys.* 4.4 as a kind of excursus, triggered by Aristotle's rather unsatisfactory arguments against the concept of place as a three-dimensional extension, and, to a lesser extent, for the commentary on the last part of *Phys.* 4.4 and on 4.5, where Aristotle tries to show that his own conception of place works better than its rival concepts. On the relation between the *Corollary* and the commentary proper and on the hypothesis, voiced by Verrycken (1990), that the *Corollary* is a later addition, see the Introduction, pp. 2-6. The *Corollary* has been translated separately in Furley and Wildberg (1991).

2. The infinite has been discussed in book 3; that this is a subject that the student of nature will have to deal with is argued by Aristotle at *Phys.* 3, 202b30-6; comments ad loc. by Philoponus *in Phys.* 386,28-387,15.

3. Aristotle may have been thinking specifically of Plato *Tim.* 52B, as Simplicius points out (*in Phys.* 521,24-32). Of course a true Platonist would not claim that 'there cannot exist anything that is not in a place' – think only of the transcendent Forms. Accordingly, Simplicius adds that 'Plato does not say this as accepting the argument, but because from the dream-like gaze at the material world, of which this does hold, we predicate it of all that is' (*in Phys.* 521,30-2).

4. Often remarked, i.e. by earlier commentators. On the flaw in the argument, see Themistius *in Phys.* 102,6-9; a very full analysis in Simplicius *in Phys.* 521,9-24, who makes it clear, in passing, that the argument was already discussed by Alexander of Aphrodisias.

5. Philoponus reconstructs the underlying argument as a hypothetical syllogism, as follows (for any *s*): if *s* is non-existent, then *s* is nowhere; but *s* exists (negation of the antecedent), so *s* is somewhere. This argument exhibits a formal structure (if *p*, then *q*; but not *p*; therefore not *q*) that is invalid. A correct contraposition (along the lines of the 'second indemonstrable' of Stoic logic) runs: if *p*, then *q*; but not *q*; therefore not *p*; in this case: if *s* is non-existent, then *s* is nowhere; but *s* is not nowhere (negation of the consequent), therefore *s* exists.

6. The notion of a 'proper' (*oikeios*) place which the elements by nature have, and which we might call their 'natural place' (although Aristotle himself does not use any Greek equivalent of this phrase, like Philoponus' *ho kata phusin topos* here), plays a prominent part in Aristotle's discussion of place and the explanation of the natural motions of the elements may even be said to be the driving force behind the whole theory of place in *Phys.* 4.1-5. There Aristotle repeatedly indicates that an adequate theory of place should be able to distinguish between places that are 'up' (the places of the light elements) and places that are 'down' (the places of the heavy elements) and in this sense he believes place has a role to play in any proper explanation of natural motion. Philoponus' own view, as set out in the *Corollary on Place*, is that place plays no such role and that the natural motions of the elements are due to their seeking the proper order and position that has been bestowed on them by the Demiurge.

7. The primacy of locomotion with regard to other types of change is argued by Aristotle in *Phys.* 8, 260a20-261a26.

8. According to this picture the strong conceptual link which exists between natural place and natural motion involves that one can on the one hand claim, as Philoponus here does on behalf of Aristotle, that we need a theory of place to properly describe and understand natural motion, whereas one can on the other hand use the evident existence of natural motion as a sign that place exists and has certain differentiations (on which see Philoponus' comments below 499,2-9). In his own view, as set out in the *Corollary on Place*, Philoponus will sever this strong conceptual link between natural motion and place.

9. See Aristotle *Metaph.* 3, 995a27-32: 'For those who wish to get clear of the difficulties it is advantageous to go into them thoroughly. For the subsequent certainty is a release from the previous puzzlement, and release is impossible when we do not know the knot'.

10. The Metrodorus mentioned here is probably Metrodorus of Chios, a (sceptical) follower of Democritus; so the philosophers here mentioned are all 'early' atomists.

11. In the doxographical overview at the beginning of Simplicius' *Corollary on Place* (Simp. *in Phys.* 601,23-4; see also 618,23-5) the view that place is extension *de facto* filled with body or bodies, as distinguished from the void space of the atomists, is ascribed to 'the most famous Platonists' and the Peripatetic Strato of Lampsacus. It is basically the view Philoponus will defend in his own *Corollary on Place* as well.

12. A similar categorical syllogism is reconstructed by Simplicius in his commentary on the same passage, *in Phys.* 524,21-9. In this particular case the main advantage of these scholastic formalisations of Aristotle's informal argument (see also Themistius' concise paraphrase, *in Phys.* 103,2-4) seems to be that they bring out more clearly the double conclusion that place must be both something that exists and something different from what is emplaced (two aspects which in Aristotle's text appear separately in 208b1 and 208b7-8 respectively).

13. Effectively, 'directions'. The Greek term (*diaphorai*) literally means 'differences', hence 'subdivisions', and refers to the fact that place, on the

Aristotelian view, is not isotropic (i.e. that places are not the same in kind, but differ *qua* 'directionality'): thus the place 'up' is by nature different in kind from the place 'down'.

14. The claim that place has a certain power – if taken literally and as something Aristotle was positively committed to – would mean that Aristotle denied not only that place is isotropic (see previous note), but also that it is inert. However, in Aristotle's own account the claim about the 'power' of place is merely part of an initial list of *apparent* characteristics of place. In the course of the discussion in *Phys.* 4 it becomes clear that this talk of a 'power' (*dunamis*) should not be taken at face value: Aristotle emphatically denies that place acts as any of the four familiar causes (see below, nn. 45, 46). In the present section Philoponus simply translates the apparent characteristic about the power of place in terms of its acting as a final cause. Simplicius (*in Phys.* 533,26-30) tells us that the question what causal status Aristotle was willing to accord to place is a matter which still requires further examination, 'because the commentators pass it over in an off-hand fashion'. So Philoponus did not have much to go on in interpreting our present passage. He seems to be following Themistius, who devotes a mere two lines to the whole issue (thus confirming Simplicius' judgment): 'Or {is it} like that for the sake of which? Yet in what way?' (Themistius *in Phys.* 105,11-12). For an explicit attack on such a view, see the *Corollary on Place* 581,18: 'it is quite ridiculous to say that place has any power in its own right; it is not through desire for a surface that things move, each to its proper place'. However, in the next lecture of the commentary it will become clear that Philoponus does not regard this as Aristotle's considered view and that he is aware of the fact that according to Aristotle place as such is not a cause.

15. Philoponus here addresses an alternative explanation of natural motion briefly discussed in *Cael.* 4, viz. that natural bodies move towards masses of material of the same kind. Aristotle rejects this in the following words: 'if the earth were removed to where the moon is now, separate parts of it would not move towards it, but towards the place where the whole is now' (*Cael.* 4, 310b2-5). Philoponus supports Aristotle on this point with some thought experiments showing that earth does not simply move towards other earth, wherever it may be, but always downwards at equal (i.e. straight) angles. His commentary here clearly remains within the confines of what he takes to be Aristotle's theory. On his own view of natural motion, on which see above, n. 8, natural elements in natural motion strive for the order and position bestowed upon them by the Demiurge. This means, among other things, that the natural motion of an object does not depend on the nature of the surrounding bodies, as it does in Aristotle, so that it could also occur in a void (see 632,4-634,2).

16. The example of the roof in this connection seems to have been a stock example in the commentary tradition. It is already found in Themistius (*in Phys.* 103,12-13), and is adopted by Simplicius (*in Phys.* 525,18-19).

17. cf. Aristotle *Cael.* 2.2, 284b6-286a2, where Aristotle argues that top, bottom, left and right are present in a non-relative sense in animate bodies only, and hence also in the heavens. The non-relative sense of right, left, top, bottom, front and back in animals is also more than once referred to in the biological works; see e.g. *IA* 4, 705a28-706a26.

18. There is some confusion in the commentators about the status of this argument. In Aristotle the reference to mathematical objects seems to be integrated into the second argument (208b8-24), which is about natural motions and directions in the physical world. It appears to be designed to further illustrate the point that in many contexts directions such as above and below are relative to the observer rather than natural, and thus to highlight – by contrast, so to speak – the special status of place and natural place in physics. Mathematical objects are not in a place, and in so far as they can be said to have positions, these are relative to us. Alexander, however, appears to have listed it as a separate, third, argument (Simplicius *in Phys.* 524,10-11), although Simplicius argues that the way he actually discusses the argument shows that this was a mistake (Simplicius *in Phys.* 526,22-5: 'Alexander clearly accepts that the illustration from mathematical objects is not a separate proof, but is used as a confirmation that differences of place are naturally determined, since those things that do not have naturally determined places, such as mathematical objects, are definitely not in a place'), although Simplicius too in the end seriously considers the possibility that we are dealing with a separate argument (*in Phys.* 526,25-30). Philoponus clearly does take the reference to mathematical objects as a separate argument (the third in number in his series). This he can do because he no longer treats mathematical objects merely as objects *lacking* a place proper, and in that sense contrasted with objects in the physical world, but as objects that have *their own* kind of place (viz. in thought). Their existence in the latter sense can then be seen as offering a separate supporting argument, through analogy, for the existence of physical place, on which see the next note.

19. Aristotle, in this part of the *Physics*, appears to emphasise that mathematical objects do *not* have a place *tout court*. Alexander as paraphrased by Simplicius (see previous note) seems to echo this: 'things that do not have naturally determined places, like mathematical objects, are definitely not in a place' (*in Phys.* 526,22-5). Themistius, however, though admitting that mathematical objects do not have a place by nature, claims that, in conceiving these objects and their relative positions, we 'conceive place along with them' (*in Phys.* 103,24). This conception of a 'noetic' place (as opposed to a 'real' or proper place) for mathematical objects is adopted by Philoponus, but also by Simplicius (*in Phys.* 526,8-9: 'relative place is capable of being applied to things not actually in place, in virtue of our conception of them'; Urmson's translation ad loc. is in need of correction). It allows both commentators to present this small section on mathematical objects as a separate argument in favour of the existence of (physical) place, though they each flesh out the argument in a different way. Philoponus argues through analogy: if even mathematical magnitudes, which exist in thought, have their places (albeit only in thought), then *a fortiori* objects which do exist in nature must have their places too (and in this case not merely in thought, but in nature). Simplicius rather suggests (*in Phys* 526,28-31) that the merely noetic and merely relative position of the parts of mathematical bodies is a notion derived from the natural places and positions of physical bodies, 'as things imagined are derived from things perceived'. Hence the existence of such noetic places and positions *presupposes* the existence of real places and positions.

20. Hesiod *Theogony* 116-17.

21. At 208b35 Aristotle claims that if we take Hesiod's lines seriously, place is 'that without which none of the other things can exist' – Philoponus first paraphrases 'none of the other things' literally as 'nothing else', but then cautiously qualifies the claim by adding that it pertains at least to the physical world of natural bodies. Of course for a Neoplatonist there exist many things –

the whole transcendent reality – that are not in a place. But also Aristotle himself holds that 'not everything that exists is in a place, but only the body that is capable of locomotion' (212b28-9).

22. All this, of course, on the view presupposed by Hesiod, a view to which neither Aristotle nor Philoponus are committed.

23. Supplying *mallon*.

24. The second section (lecture) of Philoponus' commentary on *Phys.* 4 runs from 504,10 up to 514,6 and covers *Phys.* 4, 209a2-209a31, in which Aristotle continues his introductory survey by listing a number of *aporiai* on the nature of place, which he claims may make us doubt in the end whether place exists at all. Nevertheless, just as Aristotle was not necessarily committed to the persuasive force of all arguments for the existence of place, so he need not be committed to the destructive force of these objections. Indeed, according to his own programmatic statement in 211a7-12 he is convinced that 'we must try to make our inquiry in such a way that [...] the problems are solved' (see Algra (1995) 170-81). This means that the problems should be shown either not to apply to the conception of place with which he ends up (this goes, it seems, for the *aporia* about the place of a point), or to be solved for that particular conception (which goes for Zeno's paradox of place). As in the previous section, Philoponus is primarily the neutral commentator, who explains and gives examples, while giving the arguments listed by Aristotle their due force and without anticipating Aristotle's or his own eventual solutions.

25. The neuter pronoun *auto* appears to refer loosely to place (*topos* is masculine in Greek) rather than to the preceding neuter *problêma*. Perhaps Philoponus actually wrote *auton*, masculine; MSS vary between *auto tou*, *autou* and *autou tou* in the line above.

26. Where we have arguments on both sides with regard to the same question (in this case: does place exist or not), they can have cogency on both sides only if at least one side uses premisses that are not true of the object at issue (in which case the arguments are still valid, though not sound).

27. This is indeed an important presupposition which hampered the acceptance of the concept of place as a three-dimensional extension in the Aristotelian tradition. There was no way in which such an extension could be fitted into the Aristotelian system of categories otherwise than as either a body (substance) or as the (not self-subsistent) quantitative property of a body.

28. Syllogisms of the first figure are those in which something is predicated of the middle term in the major, and the middle term is predicated of something else in the minor premiss (A belongs to all B (major); B belongs to all C (minor); hence A belongs to all C). In the example in the text the minor premiss is given first.

29. This 'second mode' (or the 'second indemonstrable') of Stoic hypothetical syllogistic amounts to what is nowadays known as the *modus tollendo tollens* of propositional logic (if *p*, then *q*; but not *q*; therefore, not *p*).

30. Aristotle himself (*Phys.* 4, 209a6-7) simply states that if place is a body, 'there will be two bodies in the same place', apparently taking the absurdity of the latter claim to be evident. In what follows here Philoponus offers some thought experiments designed to show why exactly this would be absurd.

31. 'The heaven in a millet-seed': the quotation is from *Phys.* 4, 221a22, though it is there applied in a different context (dealing with the sense in which things are said to be in time).

32. Accepting Vitelli's addition of *ou*.

33. All this still under the presupposition that what has three dimensions will *eo ipso* no longer be an incorporeal, but a body, on which see above, 505,8. Perhaps *heautou* should be read instead of the *heauto* and *heauton* of the manuscripts.

34. Aristotle merely says 'we have no distinction between a point and the place of a point' (209a10-11). It is not entirely clear how this brief remark should be put into the form of a proper argument. His claim may simply have been that a point-like place would not be recognisable as place *qua* underlying three-dimensional extension (the view of place as a three-dimensional extension seems to be the one here presupposed), since it is extensionless. In the absence of the defining characteristic of place, point and point-place might then be thought to collapse into one. Philoponus offers two different attempts to flesh out what Aristotle means, which rather suggest that the *emplaced* point would lose the defining characteristics of an emplaced something (viz. a body), so that the difference between point and point-place would collapse 'the other way round'. See the next two notes for the two arguments. It should be noted that Philoponus need not be thought to be committed to these reconstructed arguments as such; at any rate in his *Corollary on Place* 577,10-12 he rejects the underlying assumption that place, *qua* three-dimensional, must be a body, differing from an ordinary body only by its lack of further qualifications. Instead, he there takes place to be a self-subsistent 'space for body and dimensions alone, empty and apart from all substance and matter' (577,16-17), in addition crucially differing from the extension of the emplaced physical bodies by being immobile (577,18; on the immobility of place, see also *Corollary on Place* 563,13-14). So for him a point in space would be distinguishable from a point in a physical substance through its immobility.

However the argument about there being no distinction between a point and the place of a point should be constructed, the *aporia* as a whole is in the end harmless for Aristotle himself, for (1) in his view a line is not composed of points, nor a surface composed of lines etc. (see *Phys.* 4, 220a18-21: 'the "now" is no part of time, nor is a division of motion a part of the motion any more than points are parts of a line'; see further Sorabji (1983) 211-12), so that one cannot simply infer from the fact that points can have no place that bodies can have no place either; and (2) the *aporia* is solved anyhow, and explicitly, by his claim that a point is just not the kind of thing that is in a place (212b24). So the argument from replacement only concerns *bodies* or substances. Perhaps surprisingly, Philoponus in the end follows him in this respect: see his comments below, *in Phys.* 598,3-30,

35. This first argument claims that a three-dimensional place (or a 'place-body', *topikon sôma*) differs from the three-dimensional emplaced body in that it lacks any further properties apart from its dimensions. Yet, as soon as we retreat to two dimensions, or one dimension, or no dimension at all (i.e. the point), non-dimensional properties no longer play any role (a line is not hot or heavy), so that there is no longer any obvious difference between the dimensions of the place and those of the emplaced body, and the dimensions of the emplaced body and those of the place 'collapse' into one and the same thing. This, presumably, is why there is no difference between the point and the point's place.

36. The gist of this second argument is not entirely clear; perhaps the idea is that there is a correlation between the spatial qualifications of natural places and the qualifications in terms of heaviness and lightness of the emplaced bodies. In the case of a point-like part of an emplaced body the notions of weight and lightness make no sense, so that also the correlating point-place loses its qualification as 'up' or 'down' (and we may remember that Aristotle had claimed in ch. 1 that there being natural directions of 'up' and 'down' was the second reason to believe that place exists). However, this would at most mean that the point-place loses the quality of being (part of) a *natural* place, not that the point-place is annihilated altogether, as the ultimate conclusion here ('hence it is impossible for there to be a place of a point') suggests.

37. In speaking of 'body' here, Philoponus is apparently thinking of the simple bodies, i.e. the elements, which are constitutive of all other physical objects (see above, 497,15ff.).

38. By using *anairein* in this sense (i.e. as 'to rule out', 'to nullify', 'to show not to exist') Philoponus is rephrasing one of Aristotle's arguments against the existence of place in terms of the later sceptical tradition (as embodied e.g. in the work of Sextus Empiricus).

39. Thus, strictly speaking, lumping together categories of dialectic and of grammar; but see Ammonius *in Cat.* 11,1-5 where syllogisms are said to be composed of statements, whereas statements are composed of nouns and verbs, and nouns and verbs of syllables (which, however, no longer *per se* signify anything).

40. The presupposition here is that place, if identified with form or matter, would have to be extensionless. This means that Philoponus, in so far as matter is regarded (for form the idea seems to be pretty uncontroversial), is here talking about what he himself elsewhere calls 'prime matter' or the 'first subject', which 'without body, form or shape before being given volume (*exonkôtheisa*), receives the three dimensions and becomes three-dimensional' (Philoponus *in Cat.* 83, 14-18; see also *Contra Proclum* 11.1, 409,20-410,1). This indefinitely three-dimensional 'second subject' only then receives definite quantities, qualities etc. See Sorabji (1988) 23-4; de Haas (1997) 21-6.

41. So the presupposition of this *aporia*, viz. that whatever exists must be either an element or composed of elements, is wrong.

42. In this long meandering sentence, Philoponus connects the present argument (about place being neither an element nor composed of elements) with the next one (about place not being a cause). The original disjunction (elements or things made of elements) does not exhaustively cover all existing things, for there are also principles that are not elements. Hence he introduces 'principle' as a more general term than 'element' and a concomitant more general disjunction between principles and things derived from principles. Place is not derived from principles, as we are being told here, and the next argument shows that it is not a principle itself either.

43. The point that place certainly *appears* to be an object of striving, but on closer analysis is not, is also made by Simplicius *in Phys.* 533,19-21, whose analysis in many respects resembles that of Philoponus. The similarities may be due to the fact that they were both strongly influenced by Ammonius, but it is also very well possible that Simplicius simply used Philoponus' (earlier) commentary (despite the fact that he seems to have had a low opinion of Philoponus, on which see Hoffman (1987)).

44. Simplicius *in Phys.* 533,26-30 makes the same point slightly differently: 'But in general the end and the goal is not something in itself, but is the completion of that of which it is said to be the end. For in that way happiness and likeness to god are the end as being present in the happy and godlike man, not as self-subsistent, as place is said to be.'

45. Note that, despite what is often implied by modern commentators, Aristotle is careful enough never to refer to natural place as a final cause of natural motion. Instead, he specifies the end of natural motion as the element's *being somewhere* (*pou*), i.e. its *being in a natural place*, which he apparently sees as linked to the element's inner form or essence. See, e.g. *Phys.* 8.4, 255a24-5 ('what is potentially of a certain quality or of a certain quantity or somewhere (*pou*) is naturally movable when it contains the corresponding principle in itself etc.'), and 255b11 ('the actuality of lightness consists in the light thing being somewhere (*pou*), namely high up').

This being so, the analysis provided by Philoponus seems to be careful and accurate. It is paralleled, as indicated, by the analysis in Simplicius, who is even more explicit (*in Phys.* 533,22-5): 'If place is one thing and being in a place another [...] and if the end of bodies is, if anything, to be in this place, then place would not be the end.' On the dynamics of natural motion in Aristotle and in some of his ancient and medieval commentators, see Algra (1995) 195-221.

46. Here Philoponus resorts to the proper Aristotelian way of speaking of the end of natural motion as 'being in a place' (see the previous note) and asks whether this would still not warrant a loose way of speaking of 'place' as the end of natural motion. His answer, provided in the following lines, is that it would not, because (1) elements not only do not become their natural places, but are not even named after them, and (2) because final causes are somehow inherent in the things of which they are the end, whereas place is external. The same point is made, once again, by Simplicius *in Phys.* 533,26-30. The nuanced view Philoponus here presents of Aristotle's theory of natural motion and natural place suggests that he may not himself have regarded the view that place has a certain attractive *dunamis* (referred to in section 1, p. 19, and strongly opposed in his *Corollary on Place*, on which see above, n. 14) as Aristotle's considered view.

47. i.e. on the part of those who think a rational account of place can be given.

48. Reading *hê stigmê* with KM; alternatively 'And if, as {being} a point, it is one' (*hêi stigmê*).

49. The latter sentence suggests that point, line and surface are regarded as constituting parts of the body, without which it will not be in a place in its entirety. But this is not how Aristotle, in the end, saw things. See above, n. 34.

50. cf. Aristotle *Phys.* 4, 209b11-17 and Philoponus' own comments, below, *in Phys.* 515,24-516,26.

51. In speaking of the 'true concept' of place, 'which we naturally have', Philoponus uses the language of the Stoic theory of preconceptions – naturally acquired, true and criterial concepts. In a philosophical context such natural concepts have to be further articulated and elaborated into a theory. The point that is being made here – with a backward reference to *Phys.* 208a32-4 – is that such theories can move in different directions, due to the manifold aspects of the natural conception that is at their basis.

52. This third section, which runs from 504,10 to 514,7, covers what in modern editions is ch. 2 of *Phys.* 4 (209a31-210a13). Here Aristotle further pursues what he had already briefly indicated in his introductory section (ch. 1 in modern editions), viz. that various intuitions about place lead to various and incompatible definitions. He now works out two basic intuitions: place as an extension, and place as a surrounding container, and explores and criticises two definitions of place to which these intuitions might give rise, viz. the identification of place as form or as matter. One of the criticisms that he launches concerns the fact that both form and matter are intimately bound up with the substance to which they belong, whereas the place of a substance should be separate. Later on (ch. 4) he will accordingly add two further candidates for consideration: an *independent* surrounding container (the limit of the surrounding body) and an *independent* three-dimensional extension.

Philoponus' commentary is, once again, mainly expository, although his own voice can be heard in the way in which he rephrases Aristotle's sections on matter in terms of his own metaphysics of matter (encompassing prime matter as well as a so-called 'second substrate') and in the way in which, in typical Neoplatonic fashion, he plays down the differences between Aristotle and Plato and tones down Aristotle's critique of Plato.

53. According to the *Categories*, things 'homonymous' 'have only the name in common, but the definition of being which corresponds to the name is different' (*Cat.* 1a1-2). Place in the proper sense, on the one hand, and the 'common' (*koinos*) place on the other, differ in definition only in so far as the proper place precisely contains the emplaced body and nothing else, whereas the 'common' place contains the emplaced body, plus one or more other 'intermediate containers'.

54. The theoretical discussion, in Aristotle as well as here in Philoponus, will of course focus on place in the proper sense; thus, as is indicated here, it is only in this sense that place might be thought to be identical with form. Aristotle's distinction between place 'proper' and 'common' place is taken over in Sextus Empiricus' discussions of place (also otherwise heavily indebted to the discussion in Aristotle's *Phys.* 4) in *PH* III, 119 and *M* X, 5. Sextus speaks of the 'proper' (*kuriôs*) and 'non-proper' (*katakhrêstikôs*) use of the term.

55. Aristotle seems to use 'magnitude' for 'substance' or 'body', so that 'extension of the magnitude' seems to refer to the extension of a physical substance. Philoponus' metaphysics of matter – which is in the end partly based on this very Aristotelian passage about abstracting all properties from a sphere, *Phys.* 4, 209b9-11 (together with such passages as *Metaph.* 5, 1016a20-3 and 7, 1028b36-1029a28) – allows him to specify this 'extension of the magnitude' as the 'second substrate', i.e. as prime matter (the 'first substrate') endowed with indefinite extensions, and thus transformed into 'extended matter' or 'matter endowed with volume' (*ongkôtheisa hulê*), a term he uses here in his commentary. This 'distended matter' is an unqualified but also as yet unquantified extension (i.e. unquantified in the sense that it is an extension without any *specific* size), so that also in this specific sense it can be called the matter 'of magnitude'. See also Sorabji (1988) 23-4; de Haas (1997) 21-6. Note that earlier on, in arguing that place *cannot* be matter (508,17-20), Philoponus used 'matter' as (unextended) 'prime matter'. In the present context, however, he is trying to show in what sense place *could* in fact be thought to be matter; hence it makes sense for him here to use 'matter' in the sense of 'extended matter', for it is only extended matter that might be confused with place.

56. Plato does not literally call the receptacle the 'participant', although this is strictly speaking what Aristotle, and Philoponus in his footsteps, suggest, but *Tim.* 51A7-B1 has it that the receptacle 'participates in the most difficult sort of way in the intelligible'. So the label seems roughly accurate.

57. On the 'great-and-small' as principle in the 'unwritten doctrines', see Aristotle *Metaph.* 1, 987b18-22; 988a13-14.

58. Philoponus here seems to take recourse to a familiar exegetic strategy used by the Greek commentators in order to play down the apparent divergences between Plato and Aristotle or to neutralise Aristotle's criticisms of Plato, viz. by claiming that, in criticising Plato, Aristotle takes his words in their superficial sense (*to phainomenon*), and that if we take Plato's words in such a sense Aristotle is right, whereas if we look at their proper meaning he is not. See also Simplicius *in Cael.* 640,27-32: 'It is opportune to say again what I am accustomed to repeating, that the difference between the two philosophers is not substantial, but Aristotle often confronts the outward appearance (*to phainomenon*) of an argument which can be understood in the wrong sense and, out of consideration for those who take Plato's argument superficially, appears to contradict Plato.'

59. Philoponus is here describing the predominant (metaphysical rather than physical) perspective of Plato in the *Timaeus*, according to which the receptacle's role is primarily to be a receptacle for immanent forms to appear in, not a space for phenomenal bodies to be in and to move through. However, this is not sufficient to rescue Plato from Aristotle's criticisms, since (a) the receptacle is thus still space in the non-absolute sense of being the intrinsic extension of the bodies that are generated by the arrival of these immanent forms (which is indeed the type of concept of space Aristotle is describing and rejecting in the passages on which Philoponus is commenting here), and (2) there are even passages where Plato does in fact decribe the receptacle as a kind of space through which phenomenal bodies move. On these tensions in the *Timaeus*' account in relation to Philoponus' interpretation, see Sorabji (1988) 32-3; on the account of the receptacle in the *Timaeus* in general, including the various ways in which it has been interpreted, see Algra (1995) 76-110 and (for Aristotle's critique) 111-17.

60. The quotation is from *DA* 3, 429a27. Philoponus' point here is that Plato was in reality not talking about place in the physical sense, and that Aristotle should have recognised this, especially since he himself elsewhere (i.e. in the passage quoted from *DA*) also allowed himself to talk about place in a comparably non-physical sense.

61. Philoponus here describes the metaphysics of the *Timaeus* in the terms of the Neoplatonist metaphysics of his teacher Ammonius, with the *dêmiourgikoi logoi* being the trancendent Forms in their creative role. See Verrycken (1990b).

62. Here again, Philoponus describes Plato's theory in terms of his own 'metaphysics of matter': prime matter gets 'quantified' or 'distended', thus becoming a quantitatively still indefinite 'second substrate', representing the first opposition in the category of quantity, i.e. between great and small, and only subsequently acquiring a definite quantity. See above, n. 55.

63. Reading *en hôi* for the *eph' ho* of the MSS GKM.

64. A syllogism of the third figure is one in which something is being predicated of the middle term in both the major and the minor premiss.

65. Simplicius ad loc. (*in Phys.* 535,28-30) provides the very same conceptual distinctions: 'in respect of itself' (*kath' heauto*) is equivalent to 'primarily' (*prôtôs*); but the former is used in practice in opposition to 'incidentally' (*kata sumbebêkos*), the latter in opposition to 'in respect of something else' (*kat' allo*).

66. Plato *Tim.* 30A; the discordant motion is here identified with Philoponus' 'second substrate'.

67. On Philoponus' metaphysics of matter, see above, n. 55.

68. On Philoponus' metaphysics of matter, see above, n. 55.

69. At a later stage of his philosophical career, in his *De Aeternitate Mundi contra Proclum* of 529, Philoponus will himself defend this conception of matter as extension as a conception of *prime* matter (i.e. not as just the 'second substrate'); on which see Sorabji (1988) 23-30; De Haas (1997).

70. See above, n. 56.

71. Aristotelian syllogisms of the second figure are those in which a middle term is predicated of something else in both the major and the minor premisses. The example here given in the text is of the following form: M belongs to all X; M belongs to no Y; X belongs to no Y.

72. Homoiomerous substances are substances of which all the parts are structurally alike – for example: water or flesh – as distinguished from composite substances such as a hand or a foot.

73. Probably a reference to *Phys.* 3, 203a9, where it is said that the Forms are not anywhere.

74. Note that this response does not appear to be correct in so far as the great-and-small of the unwritten doctrines is concerned, which in Aristotle *Metaph.* 1 (987b18-22; 988a13-14) is presented as a principle of the transcendent Forms (or Form-Numbers).

75. But see above, n. 59.

76. That nothing will move towards matter and form, being things of which (i.e. in respect of which) there is no motion, was the gist of the fourth and fifth arguments; the further, derivative, claim that there will be no 'up' and 'down' in things of which there is no motion now constitutes the sixth argument. So the claim that matter and space are things 'of which there is no motion' belongs to both arguments.

77. This section covers the first half of what in modern editions is chapter 3 of book 4: 210a14-b8. Here Aristotle first enumerates the various senses of 'being in' something, and starts the discussion of the puzzle whether and in what sense something can be in itself by indicating that a thing can only be in itself in respect of something else, not in the primary sense. Within the context of *Physics* 4, the discussion of the senses of 'being in' something is relevant in respect of Aristotle's solution of Zeno's paradox of place (210b21-6); the question of whether a thing can be in itself involves dealing with the notion of being in a place 'in respect of something else', a distinction which is applied (though not the idea as such of something being in itself) to the discussion of the place of the universe as a whole at 212a31-b22. The section on the senses of 'being in' is rather straightforward and Philoponus' treatment of it is relatively concise. The section on whether and in what sense something can be in itself, is one of the most crabbed sections of *Phys.* 4. Philoponus notes the lack of structure and the obscurity of some passages, and he clearly sees it as his task to 'unpack' Aristotle's difficult text for students, for example by elaborating on the relation between the terms 'in respect of itself' (*kath' heauto*), 'in respect of something else' (*kat' allo*), 'in the primary sense' (*prôtôs*), and 'incidentally' (*kata sumbebêkos*) in 530,1-531,5. Parallels in wording and in the examples used between Philoponus' text and Themistius and Simplicius show, once again, Philoponus' indebtedness to the tradition.

78. It is clear that Philoponus here completely takes on the *persona* of Aristotle; the 'wrong conception' he here describes is in fact the conception he will himself defend in the *Corollary on Place*. On the relation between the commentary proper and the *Corollary*, see the Introduction, pp. 2-6.

79. i.e.: not dependent on a particular wrong-headed conception of place.

80. The references are to (aspects of) the puzzles discussed by Aristotle at *Phys.* 4, 209a23-5 and 209a13-17 respectively.

81. Compare Ammonius *in Cat.* 26,34-27,8: 'In something' is said in eleven ways: in a time, in a place, in a container, as a part in the whole, as a whole in its parts, as a species in a genus, as a genus in a species, as the ruled in the ruler, as form in matter, as in the end, as in a subject, for example an accident in substance' (tr. based on Cohen and Matthews (1991) 36). We find the same list, in a slightly different order, in Philoponus *in Cat.* 32,7-27. Simplicius (*in Phys.* 553,1-8) notes the differences between what Aristotle offers here and the lists of his commentators: the commentators distinguish 'in a vessel' from 'in a place', they treat 'as form in matter' as a separate sense, and they add 'being in time'. See also below, n. 87.

82. The Greek in Aristotle and Philoponus literally speaks of 'a thing being itself in itself' (*auto en heautôi*), but we have followed the convention of modern translators in simplifying the translation into 'being in itself', because in the Greek pronoun *auto* has no emphasis here. Wherever Aristotle wants to emphasise that a thing is in itself in its own right or in respect of itself (as opposed to 'in respect of something else') (*kat' allo*) he uses the words *kath' heauto* which we have translated as 'in respect of itself'. See above, pp. 47-8, and below, nn. 89 and 96.

83. Strictly speaking Aristotle's claim only seems to apply to those cases where we may use the same appellation for contained and container (as in the case of 'bottle'; compare 'I drank a bottle' and 'I broke the bottle') or for the contained and the combination of contained and container (as in the case of 'bottle of wine'; compare 'I drank a bottle of wine' and 'I bought a bottle of wine'). Probably this was also the case with the Greek examples ('jar' and 'jar of wine'), but in those cases where we do not call whole and parts by the same name, ordinary usage (in both English and Greek) does not appear to allow talk of the thing being in itself. Even in the cases where it does, as in the example of 'bottle', the claim does not seem to hold at the conceptual level, because in 'I drank a bottle' and 'I broke a bottle' the term 'bottle' is clearly used homonymously.

84. Taken at face value this would seem to be a strange claim. Bodies that are in contact have their limits coinciding in so far as they are in contact (cf. *Phys.* 5, 226b22-3), but such bodies are hardly said to be 'in' each other. However, the coinciding limits themselves are said to be 'together' (*hama*), which may count as a kind of 'being in' each other, albeit not in the sense of physical location, for the limits do not surround each other and they are arguably not the kind of objects that are in a place properly speaking. Perhaps it was this quasi-local relation of limits that Alexander had in mind when he isolated this separate sense of 'being in'.

85. Philoponus' objection – that the surfaces are part (not of the surrounding bodies, but) of the surrounding place – is also puzzling. First, touching surfaces are not necessarily *surrounding* surfaces, so the solution here offered cannot apply in all cases. Secondly, even where we are dealing with a *completely* surrounding surface, such a surface *is* the body's place, rather than part of its place. (Perhaps, however, Philoponus is envisaging a case like a cube which is surrounded by a surface consisting of six square surfaces, each of which may be said to be part of the surrounding place.) Thirdly, even so the two bodies and the two corresponding surfaces do not surround *each other*.

86. Echoing the words of Themistius *in Phys.* 108,10-11.

87. Alexander, as quoted by Simplicius *in Phys.* 552,30-2, has a slightly different explanation for the primacy of the local sense of 'being in': ' "in something" signifies "somewhere" generally, and somewhere is especially in a place'. But both explanations somehow take the local sense of 'being in' as the clearest case, with the implication that the other senses are all somehow derived, by analogy or metaphor, from this most basic sense.

88. Philoponus here follows Alexander of Aphrodisias (on Alexander's interpretation see Simplicius *in Phys.* 554,22-3, who, incidentally, rejects it).

89. In what follows Philoponus makes clear, going beyond what is in Aristotle's text, that, of the four relevant qualifications, 'in respect of itself' is contrary to 'incidentally' (meaning that for any subject (x) and any property (y), if (x) is (y) it is so either in respect of itself or incidentally), and that 'in respect to something else' is contrary to 'primarily', but that for the rest all combinations between the four qualifications can be allowed. Thus, something may have a particular property (1) primarily and in respect of itself (example: Socrates is walking); (2) primarily and incidentally (example: this white one is walking); (3) in respect of itself and in respect of something else (Socrates moves the door by means of his stick); (4) incidentally and in respect of something else (this white one moves the door by means of his stick).

90. See the previous note.

91. The Greek text literally says that Socrates is walking 'in respect of itself' (*kath' hauto*, neuter) rather than 'in respect of himself' because he is here regarded not *qua* Socrates, but *qua* living being (*zôion*, also neuter; or perhaps even *qua* moving thing, i.e. *qua kinoun* or *kinoumenon*).

92. In this sense the example of Socrates seeing is analogous to the example mentioned above of someone moving a door with his stick.

93. The arguments showing that it is impossible for a thing to be in itself primarily are discussed below, 533,20ff.

94. On this and similar attempts to 'unpack' and structure the rather crabbed account in Aristotle, see the Introduction, pp. 2-4.

95. The Greek *epaktikôs skopousin* refers to *epagôgê*, a sampling of cases or examples, often also translated (arguably rather over-technically) as 'induction'. Here the reference is to the list of senses of 'being in' that precedes.

96. The section which now follows covers the second half of what modern editions print as the third chapter of *Phys.* 4 (210b8-31). Here Aristotle shows that both a survey of examples (i.e. of the various senses in which being 'in' can be used) and reasoning show that a thing cannot be in itself in the primary sense, whether in respect of itself or incidentally, and he shows that acknowledging the different senses of 'being in' that have been enumerated allows us to solve Zeno's paradox of place. Philoponus fleshes out Aristotle's condensed argument by showing (i) that a thing cannot be in itself primarily in respect of itself, because in that case it will receive contrary definitions (viz. of place and of the emplaced), and that (ii) it cannot be in itself primarily and incidentally either, since, given the definition of the 'incidental' (dutifully set out and explained by Philoponus), the place would have to be an accidental property of, and hence coextensive with, the emplaced, so that (ii *a*) two bodies would (absurdly) interpenetrate; and (ii *b*), there being no real difference between place being an accidental property of the emplaced and place being in itself directly (i.e. in respect of itself), we would be back at objection (i). He next shows that, on the reading of Zeno's paradox suggested by Aristotle here, the solution offered will work, but that we will need a different solution if we accept the slightly different reading of this paradox suggested by Aristotle in an earlier passage (209a24-5).

97. Aristotle does not specify the examples beyond the mere reference to 'on any of the definitions', so what follows here is Philoponus' elaboration. Simplicius *in Phys.* 555,28-556,1 actually claims that (presumably by being so brief) Aristotle wants us to go through the examples from his list (and to add others) by ourselves.

98. On Philoponus' professed attempts to throw light where Aristotle is being crabbed and obscure, see above, n. 94 and text thereto.

99. In other words, X may be said to be incidentally Y, if X is an accident of Z and if Y with regard to itself belongs to Z.

100. So here we are back at the initial objection against the possibility of a thing being in itself in respect of itself: that it will be susceptible to contrary definitions.

101. Here (536,1-2), as Vitelli already noted, there appears to be a lacuna in the text as transmitted. Syntax and sense may be restored by inserting after *to pan to on pou einai* a clause <still it is true that everything that exists is in something (*all'oun alêthes to pan to on en tini einai*)> (or perhaps a version abbreviated to 'still it is in something' (*all'oun to en tini einai)* or something in between), and transposing the adjunct 'according to one of the senses of being in something' so as to follow this and precede the explanatory parenthesis 'for it is necessary that every physical object is in something but not in a place' (*pan gar to phusikon anankê en tini einai kai mê en topôi*). The omission is easily explained as a *saut du même au même*, the transposition as an attempt to find a use for the orphaned adjunct.

102. Here we disagree with the translation offered by Hussey (1983) 25: 'the jar will still contain the wine, not *qua* itself being wine, but *qua* a jar' (see also Waterfield and Bostock (1996) 84 for a similar translation), in which '*qua* a jar' appears to be an inaccurate rendering of *hêi ekeinos* ('in so far as the other thing is (sc.: wine)'), though arguably correct *ad sententiam*.

103. Reading (*en*) *autô*, i.e. *autôi*, with G.

104. See also above, p. 52: 'What is the fundamental difference between saying that the jar is in itself, or saying that it is a property of the wine, which is in the jar in respect of itself?' The equation seems questionable, however, for (a) as a property of the wine, the jar will not admit of the same definition as the wine; and (b) it will not be 'in' the wine in the same (local) sense in which the wine is in the jar.

105. The MSS of the *Physics* read *tini* as a (paroxytone) interrogative pronoun ('what will it be in?') rather than as an (enclitic) indefinite pronoun ('It will be in something'), but it is clear from his comments here that Philoponus read it in the latter way (as indeed did Themistius and Simplicius), which gives a more smoothly running Greek sentence anyway. However, Eudemus (as quoted by Simplicius *in Phys.* 563,17-20) seems to have read the paradox, as rendered by Aristotle, as a question; and see Morison (2002) 82 n. 7 (with n. 5 for a parallel case in *Phys.* 4, 209a23-5).

106. This is the version of the paradox which Philoponus has discussed earlier, at p. 510,3-6.

107. On place in the 'primary' or 'proper' sense, see above, n. 54 (and text thereto), and below, n. 113.

108. The lecture that follows covers 210b32-211b5, i.e. a first section of what in modern editions is ch. 4. In this more or less prefatory section (the real discussion of the proper definition of place only starts at 211b5) Aristotle lists some *prima facie* characteristics of place, states his 'research programme' on place, links the conception of place to the conception of locomotion and appends some notes on real *versus* incidental motion, and on the difference between being in a place and being in a whole. Where especially Aristotle's treatment of the latter topics may read as sets of more or less disjointed notes (see, e.g. Hussey (1983) 112-13, who speaks of 211a12-23 as 'preliminary points about change in respect of place' and remarks that it is not clear exactly how the distinction made between being in a place and being in a whole relates to what precedes), Philoponus' paraphrasing exegesis is remarkable for its repeated attempts to make clear how the various subjects discussed may be thought to interrelate.

109. The term 'common conception' (*koinê ennoia*) has its origin in Stoic epistemology. It denotes a conception that all human beings form naturally on the basis of experience. Common conceptions may be rather inchoate, in which case it falls to philosophy to articulate and refine them.

110. The explanation here referred to by Philoponus is Aristotle's way of establishing a link between (our conception of) place and the phenomenon of locomotion; the 'theorems that follow from it' are, presumably, the distinctions made by Aristotle between various *types* of locomotion and between being in a place and being in a whole.

111. This programme is carried out by Aristotle in the sections that follow of chapter 4 and in chapter 5. Philoponus will accordingly comment on its several elements in his next lectures: on the proper account of place in his lecture on 211b5ff. (546,25-552,13); on the cause of the difficulty of the subject and on the fact that the *prima facie* characteristics do indeed apply in the lecture on 212a7ff. (585,5-589,26), which comes directly after the *Corollary*; and on the fact that the puzzles are solved in his lecture on 212a31ff. (593,11-600,24).

112. Themistius *in Phys.* 111,6-17.

113. 'Primary place' (*prôtos topos*), is what Aristotle elsewhere (*Phys.* 4, 209a31) refers to as a thing's 'proper' (*idios*) place, i.e. the place 'in which a body is primarily', here explained by Philoponus in the next sentence as 'the immediately enclosing proper place'. It is the place which precisely contains the object, and nothing but the object.

114. The qualification that 'bodies' should here be taken in the sense of parts of the cosmos, is necessary, because the body of the cosmos as a whole is not in a place, according to Aristotle, because it is not surrounded by anything (*Phys.* 4, 212b22).

115. The (impossibility of there being a) void is discussed by Aristotle in *Phys.* 4, 6-9.

116. The two readings are also discussed in Simplicius *in Phys.* 565,22-7.

117. Strictly speaking, according to the Aristotelian account of place, there is no such thing as 'place as a whole' or space, but there are only individual places of individual substances. Philoponus may here be speaking with his own preferred conception of place as a three-dimensional extension at the back of his mind. Simplicius *in Phys.* 565,27-30 offers an easier and more genuinely Aristotelian solution to the same exegetical problem, by suggesting that 'above and below' here should be taken distributively as 'above or below'.

118. This is a reference to *Phys.* 4, 212a14-21. Simplicius specifies the 'exegetes' as the early Peripatetics Theophrastus and Eudemus. For their discussions of Aristotle's account in general, and of the problem of immobility in particular, see Sorabji (1988) 186-201 and Algra (1995) 192-260, esp. 222-58. Note that both Themistius (*in Phys* 111,15) and Simplicius (*in Phys* 566,12-18) add a sixth characteristic of place: that bodies naturally move towards their natural places and naturally remain in them. Philoponus later on (543,6-12) explains that he sees this not as a separate point but as something intimately linked up with place having 'above' and 'below'.

119. Below (543,16-18), i.e. in his comments *ad locum*, Philoponus speaks of a 'very useful criterion on how we should formulate our theories about things'. The passage may indeed be read as a programmatic statement of Aristotle's dialectical method; on which see Algra (1995) 173-81.

120. The terminology used by Aristotle to refer to change and motion can be confusing. The term *kinêsis* just by itself can denote change in general (including e.g. qualitative change), but also, more specifically: motion. In the present context, however, Aristotle speaks of the general category of change (*kinêsis*) with respect to place; this is then divided into two species: (1) locomotion (*phora*), i.e. change from place to place; this occurs when a body exchanges place A for place B by moving from A to B; (2) increase and decrease; this occurs when a body exchanges place A for place B by growing or shrinking into place B. The revolution (*en kuklôi phora*) of the heavenly spheres does not appear to fit in easily. On the one hand, it is loosely called *phora* and at 211a14 the spheres are said to be 'always in motion' (*aei in kinêsei*; see also *Cael.* 268b, where Aristotle distinguishes between 'straight', 'circular' and 'mixed' locomotion). On the other hand, it does not fall under either of the two species mentioned here, and hence, in a stricter sense, may not count as a form of change with regard to place. For in their revolutions the heavenly spheres do not exchange one place for another, nor do they grown or shrink. See Philoponus' comments at 556,29-557,1: 'For motion in straight lines is locomotion, but circular motion is not locomotion (*phora*), but revolution'. Alexander of Aphrodisias appears to have suggested that rotation should be regarded as a separate species of change, next to change of place (Simplicius *in Phys.* 595,20-3).

121. This is not what Aristotle actually says. Aristotle says (211a13-14) that 'the main reason (*malista*) why we think that even the heavens are in place is that they are always (*aei*) in motion'; but he is not committed to this view himself, for according to his own account, in the end, the heavens, taken as a whole, are not in a place. Philoponus is here following the exegesis of Themistius who changed *aei* ('always') in the second clause into *malista* ('mainly' or 'in particular'). See Todd (2003) 81 n. 102.

122. Philoponus here combines two different perspectives. He first states that things that increase or decrease do not exchange one place for another as a whole, meaning that they do not exhibit *phora* as a whole (i.e. motion from place A to place B, on which see above, n. 120), though their parts may indeed move from one place to another in the process of growth or diminution. He then says that they do indeed change place, in the sense that they occupy a larger or a smaller place, though not in the sense of *phora*.

123. Probably a reference to Aristotle *Phys.* 4, 212b29-213a5, where Aristotle himself draws an analogy between a thing's being in its natural place and a part's being in a whole. Philoponus seems to give his own twist to this account when in his *Corollary on Place* he explains the natural motions exhibited by the elements as their 'desiring the station which the Creator allotted to them' (581,27-8). See further his commentary below 599,10-600,25.

124. Two ways of labelling the close connection between the directionality of places (their having 'below and above') and the natural motions of the elements: the latter either follows from the former or the other way round. Philoponus here seems to gravitate towards stating the priority of natural motions (the natural impulses bodies have to move in a certain way when dislocated; see the next sentence).

125. On the relation between change with respect to place, locomotion and increase and decrease, see above, n. 120.

126. Reading *all' haptetai monon diêirêmena* with M.

127. The Greek here (and elsewhere in Aristotle and Philoponus) literally says that the limits of things that are in contact 'are in the same', which cannot as such be rendered in English; it might in principle be translated as 'in the same place', but in the present context this would be awkward, because (a) in this context place itself is what Aristotle is still looking for, so he should not be taken to use a fully fledged conception of place in the course of an argument meant to establish what place is; and (b) on Aristotle's eventual conception of place, limits and surfaces are not the sort of things that are in place (in a physical sense). Hence, to avoid the term 'place' in connection with limits that are 'together' or 'in the same' we have used the verb 'to coincide' in our translation, here and elsewhere. See also Simplicius *in Phys.* 569,35-9: 'But if he were to say "in the same" as meaning "in a place" he would clearly be speaking incidentally (*kata sumbebêkos*). For the surfaces are not in a place as such'. He adds (570,1-8) that Alexander wanted to add this special way of being 'in something' in the sense of being 'together' or 'coinciding' as a separate item to Aristotle's list in chapter 3.

128. On the difference between being 'continuous' (*sunekhes*) and 'being in contact' (*haptomenon*), see above, n. 84.

129. This whole passage in Aristotle (211b1-5) was labelled as a duplicate by Alexander and omitted in Aspasius' commentary (Simplicius *in Phys.* 570,28-34 and 571,9-10)

130. The lecture which starts here covers the second and central section of what in modern editions is ch. 4, i.e. 211b5-212a7, in which Aristotle sets out his fourfold division of possible conceptions of place and eliminates three of the four candidates. Aristotle had not bothered to show that the division is exhaustive (Hussey (1983) 115), but Philoponus offers the required proof. He next spends most of his commentary on Aristotle's arguments against the conception of place as a three-dimensional extension, which he claims are very obscure and have been variously interpreted by the Aristotelian tradition (548,16-18). Nevertheless he uses that tradition – in particular a large chunk from Themistius – to explain Aristotle's text in a faithful way without criticising it. He ends the theoretical part by announcing that he will come back to this issue later on, i.e. in the *Corollary on Place*, which is inserted right after this lecture (557,8-585,4). It is there that we find the systematic critique of the Aristotelian position that is kept hidden in the present mainly exegetical lecture.

131. i.e., presumably, as opposed to the immanent form which determines the thing's essential characteristics.

132. The same exegesis of the rationale behind the fourfold division is to be found in Simplicius *in Phys.* 571,1-6.

133. For the translation 'coincides' for *en tautôi onta*, here and elsewhere in this context, see above, n. 127.

134. Note that Philoponus treats the arguments against the identification of place as matter or form as 'compelling'. This is not the case for the arguments against place as an extension, which he discusses next, and which he will refute in the subsequent *Corollary on Place*.

135. The reference must be to two passages in Aristotle's discussion of the void: *Phys* 4.8, 214b24-7, where it is asked how a body can be in void, i.e. in a separate extension: even if it is in this void as a whole 'the part, if not placed separately will not be in place but in the whole'; so there as well the conception of all the parts of the body being in place is adduced as an absurdity; and 4.9, 216a26-b20 which focuses on (the alleged absurdity of) various extensions interpenetrating and on the fact that positing a self-subsistent extension next to the extension of individual substances replacing each other is superfluous.

136. The section that starts here (548,18-549,11) is a commentary on 211b19-23, which gives Aristotle's first argument against place as an extension. The gist of the argument is that there will then be an infinity of places. Philoponus takes this claim to be based on the fact that the extension of a body is itself infinitely divisible, so that this must also apply to the separate extension that it occupies (548,26-8). There is no explicit mention here, as yet, of an infinity of *nested* places, a notion which emerges in his exegesis of Aristotle's second argument (211b23-5) at 549,11ff.

137. Strictly speaking this is not true, for the infinite divisibility of both matter and place on this view is a potential divisibility, so there is no actual infinity involved; in the *Corollary on Place* Philoponus explicitly denies that there is an actual infinity of parts in either the place or the emplaced object (557,11-560,18). There he interprets the argument he here gives as implying that place, by interpenetrating the body, will divide it into an infinity of parts. Also Simplicius (*in Phys.* 577,24-578,14) rejects the argument about the actually infinite number of places, as he also rejects the second argument in Aristotle (about place being in place). It is not obvious, however, that it is the threat of an actual infinity that was bothering Aristotle here in the first place. He rather seems to baulk at the idea itself that the parts of a substance – however many – should be thought to be in a place in their own right (see 211b25: 'but the place of the part is not different').

138. Philoponus appears to have no trouble with the MS reading *ei de ên ti diastêma to pephukos kai menon en tôi autôi*, where Ross adopts an emendation of Laas (1863) based on what is 'perhaps' (fort.) the reading of T, *<kath' hau>to pephukos <einai> kai menon* 'naturally such as to exist on its own and remaining', and Carteron adopts Themistius' reading *pephukos einai kai menein* 'naturally such as to be and to remain'.

139. Following Vitelli's punctuation, which takes *en tôi autôi* to go with *diastêma ... menon* rather than with what follows (*apeiroi an êsan topôi*). On the latter punctuation one would have to translate 'then there would be an infinity of places in the same spot'. In its turn, the translation 'in the same spot' would refer to an infinity of *nested* places, which may in fact have been what Aristotle had in mind (see Morison (2002) 124-5), but does not appear to be what Philoponus is thinking of here.

140. Providing an exegetical paraphrase of the last sentence of the quotation from Aristotle.

141. What follows is an exegesis of Aristotle's even more condensed second argument at 211b23-5: 'At the same time place will be moving, so that there will be another place of place, and many places will coincide'. Here Philoponus, following the exegetical tradition (see the extensive paraphrase of Themistius which follows at 550,9ff.), does assume that Aristotle was thinking of coinciding *nested* places; for example in the case of a jar: the place of the jar of water as a whole and the place of the water that it contains. In truth, on a conception of place as a *self-subsistent* three-dimensional container, these are merely overlapping sections of one and the same space or place. Here, however, they are taken as nested, because the place of the water is conceived of as definitionally tied up with the inner surface of the jar (the extension in between its inner limits), so that it is no longer a part of an absolute place or space, but an extension moving along with the jar when that changes place. Hence, absurdly, we have a place in a place (the place of the water in the place of the whole jar) as well as a moving place (when the jar as a whole is moved, the place of the water moves along with it).

142. So we have three co-extensive and coinciding extensions: (1) the extension of the water itself (being a quantitative property of the water), (2) the place of the water taken as the extension in between the hollow surface of the jar (which thus defined is taken to move along with the jar), and (3) the water-sized part of the place in which the whole jar has come to be (and which may be taken as 'absolute', i.e. independent of the moving jar itself). In truth, on the conception of place as extension (2) and (3) would count as two different specifications of one and the same part of the same, independent, extension.

143. Compare Aristotle *Phys* 4.8, 216b9-12: 'And if there are two such things, why should there not be any number of things coinciding?'

144. The following example is designed to illustrate the idea of nested places.

145. What follows (550,9-551,20) is a paraphrasing rendering of Themistius *in Phys.* 116,12-117,5. Themistius' text in its turn is an expanding paraphrase of Aristotle's first argument (i.e. that we get an infinity of places) – expanding it mainly in the sense that it (a) explicitly adds the idea of nested places and (b) makes clear that the absurdities resulting from the idea that the parts of substances are in a place in their own right do not follow on the Aristotelian account, where these parts have no place of their own (thus making sense of 211b25-6).

146. The list of partly coinciding extensions here given – using the example of a vessel of water which is being moved to a new place distinguishes between: (1) the extension occupied by the water in the vessel, defined as the place between the inner limits of the vessel, and hence as a kind of 'gap' moving along with the vessel; (2) the parts of (1); (3) the independent place in which the vessel comes to rest after having moved; (4) the part of (3) that is occupied by the water alone; (5) the parts of (4); (6) the extension of the vessel itself, *qua* body; (7) the extension of the water itself *qua* body; (8) the parts of (6); and (9) the parts of (7). However, according to the view that place is an independent three-dimensional extension, (1)-(5) are not separate and partly overlapping extensions, but partly overlapping parts of one independent extension. Moreover, the fact that (6) and (7) coincide with parts of (3) is on this view entirely unproblematic. So the argument as a whole is misguided and stems from a failure to see what it really means to have an independent (absolute) three-dimensional extension underlying the changes and motions of extended substances. Note that Philoponus freely moves between the use of the words *aggeion* and *amphoreus*, which we have tried to cover as 'vessel' and 'jar' respectively.

147. In the *Corollary on Place* (560,18-563,25, esp. 563,7-14) Philoponus will argue, against arguments of this kind, that (1) on the view here attacked there are in fact not infinitely many, but just two extensions that coincide, viz. the extension of place and the extension of the emplaced body; and (2) that even an infinite number of coinciding extensions is unproblematic as long as these extensions are all incorporeal (for if any number of lines or surfaces can coincide, why not also any number of incorporeal three-dimensional extensions?).

148. In this section Themistius explains why the allegedly absurd consequences that follow for those who claim that place is an extension do not follow for those who define it as the limit of the container (i.e. for the Aristotelian view). *Allegedly* absurd, for in his comments on 212a31ff. (chapter 5 in modern editions) below, Philoponus explicitly rejects the suggestion that adopting the conception of place as a three-dimensional extension commits one to the view that there must be places of parts, or of surfaces, lines and points (598,18-30). See Introduction, p. 9.

149. This is an expanded version of Themistius *in Phys*. 117,5-8. Themistius' argument is, once again, unsound, for according to the account that is being attacked, the place of what is in between the limits of the vessel will not be *in* the place of the whole vessel as in a place, but it will be a *part* of it.

150. What follows is an imaginary defence which those who think that place is an extension might (but, in so far as we can see, did not in fact) put forward in order to dodge the conclusion that many places will coincide. If they should think that places lose their existence, as a separate extension, once they are filled, a *reductio ad absurdum* will follow, for then there will never be an independent extension, for the extension has always been, and will always be, occupied.

151. The reference is to the *Corollary on Place*, which is inserted right after the last comment of the present lecture (on the *lexis* of 212a6) and runs from 557,8 to 585,4. In it Philoponus presents us with systematic observations, not directly keyed to Aristotle's text and its exegesis (hence here called 'external arguments'), on the weaknesses of Aristotle's own account and on the merits of the conception of place as a three-dimensional extension. The *Corollary* has been translated separately in Furley and Wildberg (1991).

152. For the Greek term *en tautôi*, here translated as 'coincide', see above, n. 127. The fact that the limits of place and of the emplaced body coincide is here presented as giving rise to the mistaken view that the limit of the emplaced body – i.e. form – is place.

153. *ti tukhon metempiptei sôma* ('ever different bodies replace each other') here is a loose paraphrase of Aristotle's *to tukhon empiptei sôma* in 211b18.

154. Following the punctuation of Ross' edition of Aristotle, rather than of Vitelli's edition of Philoponus, which takes *en tôi autôi* to go with *menon* ('remaining static in the same place').

155. In other words, 'that exists naturally' and 'that remains static' are added to distinguish the extension of physical place from the extension of geometrical entities on the one hand and from the extension of physical bodies on the other.

156. This, of course, is not true, although it basically seems to be what Aristotle had in mind; on which see above, n. 141.

157 On these three extensions, see above, n. 142.

158. It is a reply in the sense that it shows that the absurdities that follow for the view that place is an extension do not follow for the view endorsed by Aristotle himself.

159. That is: as a part it is only incidentally in the place of the whole.

160. Even if the water, *qua* part of the whole, moves along with the vessel, it does not change place in itself, or in respect of itself. The Aristotelian account thus works out the common sense notion that the water, while moving, remains *in* the vessel; and what for the account of place as an extension, and perhaps for any coherent physical account of place and motion, would be the more fundamental phenomenon (viz. the water's changing place along with the vessel) is downgraded to incidental motion and to acquiring a new place only incidentally. Note that later on Aristotle stresses the immobility of place (as distinguished from a vessel) and speaks of a vessel as a portable or mobile place (212a14-16); in the present example, however, the required immobility of place (the inner limit of the vessel) is either disregarded or fleshed out in a rather minimal and unhelpful way, viz. only with respect to the emplaced body itself.

161. Philoponus reads the words 'not in the place in which they come to be, which is part of the place which is the place of the whole cosmos' as if they refer to Aristotle's own account; but as such they do not make sense (for there is no 'cosmic place' according to Aristotle; see Hussey (1983) 116, who finds this remark 'impossible to accommodate'). However, Aristotle probably intended them to refer to the conception of place as extension which he is criticising: air and water (and their parts) replace each other in one and the same Aristotelian place, but *not* in an extra, independent three-dimensional extension which is part of the extension which is the place of the whole world. See Simplicius *in Phys.* 576,17-23 for a similar interpretation. See also Morison (2002) 126.

162. Reading *ean* instead of *hina* of the MSS.

163. We are dealing with a body that is quasi-continuous (though, as the next sentence indicates, not really continuous) with its surroundings and that does not move away from them (and in that sense is 'not separated').

164. In other words, it is in virtue of the fact that matter and place are both a 'that-in-which', that matter may be wrongly taken to be place, especially in contexts where there is no locomotion (for then the difference between the two, viz. that matter is the 'that-in-which' for forms, and place for mobile substances, is least evident). On Aristotle's critique of this conception of place as endorsed by Plato, see Algra (1995) 110-18.

165. i.e. it is not able to be at times moving, but at other times not.

166. The reason why the heavenly sphere cannot exhibit locomotion is also the reason why it is not as a whole in a place, viz. that there is nothing outside it.

167. The lecture which starts here, and which follows immediately after the *Corollary on Place*, covers the last part (212a7-30) of what in modern editions is chapter 4. In it Aristotle discusses the cause of the difficulty of the subject, and the difference between a vessel and a place (and in that connection the required immobility of place, partly with the help of the notorious river-example); he ends with some rather sketchy notes that may serve to show that the resulting definition of place as the 'first immobile limit of that which contains' (212a20) fits a number of the characteristics that belong to place according to the common conception. Most of the theoretical part of Philoponus' lecture (586,25-589,26) is devoted to these notes. Philoponus works them out into a neat list and takes the opportunity, clearly building on the findings of the preceding *Corollary*, to show that the conception of place as a three-dimensional extension fits the said elements of the common conception equally well, if not better. Whereas in the earlier sections of the commentary on *Phys.* 4 (i.e. the sections which preceded the *Corollary*) criticism of Aristotle was, when presented at all, confined to the obscurity of some passages, this lecture (as well as the next, which covers what is chapter 5 in modern editions) also contains critique of the *contents* of Aristotle's account. The intervening *Corollary* thus appears to serve as a critical turning point in the linear structure of the commentary.

168. Both the point about the instantaneous nature of replacement and the point made here about air not provoking sensation like the other elements seem to derive from Themistius' exegetic summary of the relevant section at *in Phys.* 118,11-23.

169. In line 2 the word *apodexamenês* printed by Vitelli must be a mistake for *apodexamenon*.

170. Philoponus clearly reads Aristotle's example of a boat in a river as involving a boat moving along with the flowing water, which makes the case analogous to that of water in a vessel that is being moved. For other variant versions of the problem (e.g. a moored boat with the current flowing by), see Sorabji (1988) 188-92. For the general problem of the immobility of place, and the way it was discussed in the early Peripatos and in the later Aristotelian tradition, see Algra (1995) 222-58.

171. In taking 'the river' here as 'the river banks', and in adding the qualification that this place is not the primary place, Philoponus is following the interpretation of Alexander of Aphrodisias (as rendered by Simplicius *in Phys.* 584,5-16).

172. As the next sentence makes clear, 'the centre of the universe' here does not refer to the geometrical centre, which is a point and as such neither a place nor in place, but to what is *at* the centre in a less strict sense, i.e. the earth (*qua* element). Aristotle claims that 'the containing limit which is towards the centre is below, and so is the centre itself' (212a25-6). In other words, both 'centre' and 'below' can refer both to what *is* at the centre and to the central *place* (i.e. the place containing what is at the centre). See also below, 587,16-21.

173. Or perhaps *ametablêtos* and *akinêtos* (here both translated as 'immobile') should both be translated as 'unchangeable' in which case the last phrase would read 'it is nevertheless unchangeable as a whole according to every type of change'). Yet that would perhaps be claiming too much; after all, rotation should either count as a separate kind of change, or as a subspecies of motion. Moreover, in the present context it would appear to be merely motion (or rather immobility) that is at stake. However, there seems to have been some confusion in the commentary tradition, to judge from Simplicius *in Phys.* 585,17-20: 'Still as a whole it does not exchange place for place, but remains always retaining its unchanged place, which is what the argument is especially concerned to show, rather than its unchanging existence.'

174. Philoponus does not here define what he means by 'formally' and 'numerically', but his meaning is more or less clear. The distinction is between two senses in which a body (or a surface of a body) that is immobile can be said to remain the *same*. The inner surface of the sphere of the heavens remains the same formally in so far as it always remains in the same position with regard to the rest of the cosmos, and numerically in so far as it is the surface *of the same body*. The inner surface containing the earth also remains the same formally in so far as it always remains in the same position with regard to the rest of the cosmos, but it does not remain the same numerically (for it is constituted by the ever fleeting and interchanging surfaces of water and air). The distinction found its way in the mediaeval Latin commentary tradition as the distinction between 'formal place' and 'material place', on which see Grant (1981) 63-72.

175. The 'parts of the water' should here not be taken in the sense of parts of water as a continuous mass, but as the separate 'waters' (i.e. ponds, rivers, lakes, seas) on the earth.

176. On the types of motion, see also nn. 173 and 204.

177. Aristotle does not say so explicitly. The point is at most implied in the phrase 'the containing limit which is towards the centre is below, and so is the centre itself' (212a25-6) and in his use of 'above' (*anô*) and 'below' (*katô*) as labels that can be used for both places and emplaced bodies.

178. The same interpretation of 'above' in Themistius *in Phys.* 119,10-11: 'But what creates the upward place is the extremity of the circular motion, and anything related to it'; and Simplicius *in Phys.* 585,20-1: 'One must note that he says that the above is the concave surface of the lunar sphere, so that what changes by locomotion is the sublunar'. For a different interpretation by Porphyry, see below, n. 193.

179. The same point is made in the *Corollary on Place*, 564,4-14. As a criticism it goes back to Theophrastus' first *aporia* against Aristotle's concept of place (Simplicius *in Phys.* 604,5-8; fr. 146 FHSG).

180. This argument as well also appears in the *Corollary on Place*, 564,14-32.

181. The gist of this objection seems to be that since place is the limit of the surrounding body, it is the limit of a *particular* substance and that hence when this substance is replaced by a completely different one, the notion of formal identity becomes hard to make sense of. Indeed, in the case of the example adduced, the body at issue would no longer be in the water, but in the air. On Aristotle's own principles (according to which a body is located with respect to its immediate surroundings), this would surely mean that it changes its place *tout court*. What is more, Aristotle's account of natural motion presupposes that it makes a crucial difference by *what kind of* body an object is surrounded.

182. This may be reference to an attempt on the part of unnamed Aristotelians to defend the immobility of place in terms of a relation between place and its surroundings. But if that is what it is (rather than Philoponus drawing his own conclusions from the example at 211b25ff. of water in a moving vessel remaining in the same place), Philoponus seems to have mistaken the gist of it. For such a defence works only if one considers the (fixed) relation between place and surroundings *that are themselves immobile*, such as the poles or the inner surface of the sphere of the heavens, which is indeed what was suggested by Aristotle's pupil Eudemus of Rhodes (Simplicius *in Phys.* 595,5-8) and by mediaeval scholastic commentators who used such relations to underpin their conception of 'formal' place. See Algra (1995) 252-3 on Eudemus, and Grant (1981) 63-72, with nn. 20 and 21, on the mediaeval versions. However, Philoponus here specifies the relation at issue as one between place *and the emplaced thing*. Given that place is always what immediately surrounds, that relation is in all cases the same, as he notes, and it does not help to provide the required immobility either.

183. The same strong conclusion on the impossibility of Aristotelian places being immobile can be found in the *Corollary on Place* 564,29-32.

184. In other words, Philoponus argues, the surface only serves as a place in so far as it is considered as a physical entity, and hence involved in processes of change and locomotion (the term 'incorporeal, as applied to a surface, is in itself of Stoic provenance, but the point of the anonymous Aristotelians here seems to be the Aristotelian one that it is not part of the physical world which consists of mobile and changeable substances). Place as a surface, in short, should not be considered as an incorporeal part of an independent 'grid' of locations, but as the incorporeal limit of a corporeal substance. In his comments ad loc. (590,15-20) Philoponus provides yet another Aristotelian defence, without countering it, on which see below, n. 198.

185. In this passage Philoponus makes clear that he takes the term 'containing' specifically and exclusively in the (Aristotelian) sense of 'surrounding', and he accordingly tries to save the idea that also place taken as an extension in this sense 'contains' what is emplaced, by emphasising that a body's three-dimensional place includes its own 'surrounding' boundary. It might be argued, however, that 'to contain' can also mean 'to receive' or 'to be occupied by' and that also in this sense place as an extension may be said to 'contain' the emplaced body. On the difference between the 'circumscriptive' and the 'receptive' sense of 'containing', see Hussey (1983) 108 and Morison (2002) 58-60.

186. So an object that is not literally in a vessel, but emplaced in surroundings by which it is carried along as if it were carried along in a vessel, is in these surroundings as in a vessel. Of course this raises the question what then is the immobile place (as distinguished from the mobile vessel) of such an object, which Aristotle answers by his enigmatic and arguably unhelpful 'the whole river'.

187. Literally: 'means/wants (*bouletai*) to be immobile'.

188. Given the fact that the *theôria* has stressed the fact that Aristotelian places cannot really be immobile, it is surprising that Philoponus here (in his exegesis of the *lexis*) presents an Aristotelian defence without countering it. The defence is indeed rather lame, for it merely makes the *ontological* point that a surface cannot move in its own right, whereas the immobility required is of course of a *physical* kind: what is at stake is whether place moves *at all*, be it in its own right or incidentally. So Philoponus could have countered this defence along the same lines he used in countering the argument (588,20-4) that place as a surface is immobile *qua* being incorporeal (on which see above, n. 184).

189. Note that we are not following Vitelli in reading *kai*, with M, in line 2. It may be explained as being due to a loose contamination between or contraction of 'for these *too* have a natural place they move toward, and this is the centre'. And it is omitted in G.

190. As has been explained in the *theôria* (587,12-16), the place of the earth, though formally always the same, is materially (or numerically) constituted by ever different surfaces of air and water, according as the waters on the earth (and the air moving around it) come to be and perish and change positions.

191. One substance, therefore without numerical difference.

192. Reading *epeidê esti phêsi* in line 14, with G.

193. Note that in the former case 'extreme part' (*eskhaton*) refers to the inner surface of the heavens, whereas 'the extreme part itself' refers to the extreme part of the sublunary world, i.e. the region of fire. Simplicius (*in Phys.* 587,10-12) records an attempt by Porphyrius to provide these two 'extremes' with the same reference; 'the above [...] is either the limit of the extreme, as of the aether, which is its place, or it is the extreme itself, which is the aether'. But this is unlikely to be correct as an interpretation, because (1) the limit of the aether can hardly be the place of the aether, and (2) Aristotle connects 'above' and 'below' to the rectilinear motions of the elements in the sublunary world, so that 'above' and 'the extreme itself' naturally refer to the end of upward motion, i.e. the sublunary region of the light.

194. The 'theoretical introduction' (*theôria*) is the general overview of the argument of a whole stretch of Aristotle's text with which each lecture starts out, and which is followed by the actual *lexis*, the explanation of the text (often down to the level of the meaning of individual words) keyed to individual passages. The reference to the *theôria* here is to 587,16-21.

195. This is an expanding paraphrase of Themistius *in Phys.* 119,5-12. The difference between his explanation and the preceding one is that Themistius defines the area of the 'below' as not just the earth, but as earth, water and air.

196. i.e. not the earth.

197. Apparently 'the same thing' here refers to the common surface between any two bodies that are in contact. However, on Aristotle's considered view, although the surfaces of body A and body B coincide if and in so far as they are in contact, the surface of body A is nevertheless formally and definitionally different from the surface of body B.

198. That place is void in its own right, but always *de facto* filled, is what Philoponus has defended in the *Corollary on Place* 569,8-10.

199. Philoponus here simply explains the Aristotelian position without criticising. In the *Corollary on Place* he argues, by contrast, that place as a surface, is *not* equal to what is in place (564,4-13).

200. This last section of Philoponus' commentary on Aristotle's discussion of place in *Phys.* 4 covers what is in modern editions chapter five of Aristotle's text. This chapter roughly consists of two parts: 212a31-b22 deals with the question whether and to what extent the heavens and the cosmos as a whole are in a place; 212b22-213a11 then finally shows that the puzzles that were raised with respect to place can be solved on Aristotle's account, and that the phenomenon of natural motion in connection with natural places can be accounted for. The text of 212a31-b22 is condensed and difficult, and matters are further complicated by the fact that Aristotle designates his subject as the *ouranos*, which is said not to be in a place as a whole, but to have places for its parts in so far as they move and contain each other (hence, they somehow act as each other's places). However, the word *ouranos* can refer either to (1) the whole cosmos (as a synonym of 'the universe' or *to pan*), or to (2) the outer sphere of the heavens, or to (3) the heavens as a whole. Interpretations of what Aristotle says (and especially of what he means by 'the parts' of the *ouranos*) naturally differ according as one opts for (1), (2) or (3). Waterfield and Bostock (1996) opt for (1) and take the whole of 212a31-b22 to be about the (place of the) universe. Hussey (1983) 119 rather assumes that Aristotle is moving between the different senses of *ouranos*, as indeed does Philoponus in the various sections of his commentary. Simplicius *in Phys.* 594,35-37 actually complains that 'it is clear that he was calling either the whole universe or the whole of that which revolves "the heavens"; but he created much unclarity in the passage before us by saying sometimes "the heavens" and sometimes "the universe" '.

Philoponus does not attempt to come up with a single, unified interpretation of this first part of Aristotle's chapter. He basically sticks to a presentation of the most common interpretative options in the exegetical tradition, although he does not offer a sustained critique of the Aristotelian position either. But perhaps the reader is expected to keep in mind what has been said in the critical *Corollary on Place* (565,12-21): 'Hence, when they try to explain how the sphere of fixed stars could move in place when it is not in place, they throw everything into confusion rather than saying anything clear and persuasive. For they cannot deny that the sphere moves in place, because they cannot even make up a story about what {other} kind of motion it would have. However, they cannot explain what is the place in respect of which it moves, but like people playing dice they throw out first one account, then another, and through them all they destroy their original assumptions and agreements. For by concealing the weakness of his account with obscurity, Aristotle licensed those who want to change their stories however they wish'.

Philoponus' commentary on the second part of Aristotle's chapter, 212b22-213a11, contains some remarkable references to his own preferred conception of place as an independent three-dimensional extension, which is said to solve the puzzles equally well as, or even better than Aristotle's account; and it makes a surprising use of the concept of the 'force of the void' in explaining Aristotle's account.

201. At *Phys.* 4, 212a31-b22, Aristotle appears to move from the question of the place of the universe as a whole (with the added thought experiment of what would be the case if the universe were a single mass of water) to the question of the place of the heavens (*ouranos*). In his commentary here Philoponus starts out (59316-25) by considering the question of the place of the universe (*to pan*), with some added thoughts about the question of the place of the *ouranos* being analogous (594,17-18), and then goes on (593,25-594,28) to speak about the place of the outermost sphere (the sphere of the fixed stars, see below, n. 203), Later on, however, e.g. in his commentary on the *lexis* of 212b10 (602,14-24), he takes Aristotle to be talking about the heavens as a whole (i.e. the eight spheres of fixed stars, planets, sun and moon).

202. The Greek term *to pan,* here translated 'the universe', literally means 'the all', hence 'the totality' or 'everything', i.e. everything there is. So by definition *to pan* cannot have anything outside it.

203. In the lines that follow, Philoponus, in talking about 'the heavens' (*ouranos*) appears to be talking about the outermost sphere, as is indicated by the alternative explanation offered by the 'exegetes' that he appends at 594,29ff. The two explanations of the emplacement of the outer sphere (which he here ascribes to 'Aristotle' and to 'his exegetes' respectively) are summarised once again in 601,21-2.

204. Rotation constitutes a problem for Aristotle's account of place, because, though commonly conceived as a species of motion, it cannot be explained in terms of the moving object's changing place, if place is a containing surface. At most it can be explained in terms of the *parts* of the rotating body changing their *potential* places, on which see the next footnote. All this may be the reason why Aristotle sometimes seems prepared to include rotation as a species of locomotion (e.g. *Cael.* 268b17), whereas at other times he uses a more restrictive concept of change with respect to place, which includes only locomotion in the sense of straight motion, as well as increase and decrease; on which see also above, n. 120. Alexander of Aphrodisias suggested that rotation should be regarded as a separate kind of change, different from change of place (Simplicius *in Phys.* 594,20-3).

205. The 'parts' of the *ouranos* that Philoponus mentions here are, presumably, segments of the outer sphere, which can indeed be said in a sense to exchange their places in the process of rotation. The problem is how to make sense of this 'exchange' of places in the process of rotation in terms of Aristotle's account of place as the limit of the surrounding body. For even if the parts of the outer sphere in a sense surround each other and 'are like places to each other', they also move along with each other in the process of rotation, so that in this sense they do not change place individually. Perhaps Aristotle is here suggesting that the continuous parts of a rotating body have *potential* places (a notion he introduces in this same context, 212b5), but if he is indeed applying the notion of *potential* places to explain the *actual* rotation of the outer sphere, he seems to be gravitating towards the idea of an independent 'grid' of containing surfaces. This appears to be what Morison (2002) 166 ascribes to him: 'each segment of the outer sphere has a specifiable place within the universe – it would make a recognisable "hole" in the universe'. But this is hardly consistent with what we may call the ontology of Aristotelian places as surfaces *of containing bodies*. Themistius remarks that the parts of the outer sphere cannot in fact be in a place potentially, because they are inseparable from the whole (*in Phys.* 121,6-7). See further Philoponus' own critique of this solution in his *Corollary on Place* 566,7-31.

206. This first interpretation of what should be considered to be the place of the parts of the outer sphere may have been adopted from Alexander of Aphrodisias (see Simplicius *in Phys.* 593,13-15).

207. Philoponus now appends a second interpretation. The exegetes (one of them for sure being Themistius: see his *in Phys.* 121,2-5) thus exploit the fact that the segments of the outer sphere are not merely surrounded by each other, but also, on the inside, by the surface of the sphere of Saturn, and this surface at least would not move along with the segments of the outer sphere. Yet as a surface it does not contain the segments of the outer sphere on all sides, and hence it would fail to meet the requirement of being equal to what is emplaced. Moreover it would itself be moving rather than immobile. Hence Philoponus here adds that this is a conception of place only by analogy. For a more sustained critique of this view of the exegetes, see Philoponus' *Corollary on Place* 567,8-29. Nevertheless, he later indicates that he regards it as 'more proper' (*kuriôteron*, 601,22) than the view which takes the parts of the outer sphere to act as places *for each other*, on which see below, n. 231.

208. i.e. the objects in the sublunary world; locomotion is here to be taken in the strict sense which excludes rotation, on which see above, nn. 120 and 204.

209. Of the two views of the soul here hinted at, the former is, roughly, the Aristotelian one, the latter belongs with the Pythagorean and Platonic traditions.

210. The puzzle, and Themistius' solution, can be found in Themistius *in Phys.* 120,21-8.

211. In other words, since it can be used in the case of what is something in respect of something else, it may be used in the case of what is something in respect of its parts, since 'in respect of its parts' would be an instance of 'in respect of something else'.

212. For this distinction, see above, 530,1-531,5.

213. Presumably the idea is that in virtue of its being a part of a continuous whole (and thus different from that whole) the part is in a place in actuality in respect of something else (viz. the whole), whereas in so far as it is itself continuous with the whole, it shares in the properties of the whole and hence is in a place in respect of itself as well.

214. Aristotle says (*Phys.* 4, 212b17-18): 'For the heavens are perhaps the universe'.

215. The puzzle about growing places and growing bodies that Aristotle had offered at 209a25-30 connected the issue with the claims that (1) every body is in a place and that (2) in every place there is some body. These things being so, how can there be growth of bodies, unless we say that place grows along with bodies? Philoponus now shows how Aristotle can maintain that bodies can grow and come to occupy larger places, without assuming either that there is void (which would be a denial of (2)), or that there are bodies which are not in a place (which would be a denial of (1)), yet also without assuming that place itself grows. Note that in speaking of 'takes up just that place' Philoponus appears to have his own notion of place as an extension at the back of his mind.

216. See *GC* 1, 321b22-322a4, where Aristotle explains that when a thing grows by the accession of food, the food accedes not to every part of the matter (which is in constant flux), but to every part of the body or body part *qua form*.

217. In principle, a conception of place or space such as the one endorsed by Philoponus could leave room for the notion of the emplacement of parts (Themistius actually argued that it necessarily *implied* such a notion, and adduced this as one of the points on which the Aristotelian account was superior; see the quotation in the text above, 551,1ff., esp. 515,10-12 and n. 148). However, Philoponus sticks to the Aristotelian substance-ontology according to which the continuous parts of a substance are not themselves movable substances and as such not the sort of things that are in a place of their own. Moreover, Philoponus wants to avoid the suggestion (which he thinks is at the basis of one of Aristotle's objections against the conception of place as extension at 211b19-23; see *Corollary on Place* 557,12-560,18) that place *qua* three-dimensional extension divides the bodies that are in place into (an infinite number of) parts that are in place in their own right: 'if the body is not divided by the extension, and the extension does not pass through the body, why should it necessarily follow that the part is in a place in itself?' (*Corollary on Place* 578,2-4).

218. On Philoponus' own conception, points, surfaces and lines *qua* limits can coincide, but this does not mean that they are in each other as in a place: see *Corollary on Place* 558,19-22: 'I am not saying that one surface will be in the other as in a place: even if it is not as in a place, nevertheless they are applied to each other [...]'. Here, however, his point is rather that even if we allow the inference from coinciding to being in a place, the argument applies to the Aristotelian account as well.

219. In other words, surface, line and point are neither place-fillers (since they do not 'fill', presumably because they are not three-dimensionally extended) nor place (i.e. void) itself. So they are 'in' each other in a non-local sense.

220. *Phys.* 4, 212b28-9.

221. i.e. he ignores the motion and location of the inner spheres, and is speaking of what is capable of locomotion in the strict sense: the bodies in the sublunary world.

222. Philoponus praises Aristotle's account of natural motion in terms of the natural relation between an element and its neighbours, no doubt because it resembles his own account of natural motion as a matter of a body's striving after the natural order and position bestowed on it by the Creator (on which see Introduction, p. 8), even if he is not himself prepared to let natural *places* play any role in the explanation of the phenomenon.

223. *GC* 1, 331a23.

224. The 'fifth element' (*aithêr*) that constitutes the heavenly bodies does not share a 'tally' with the four elements in the sublunary world (hence it cannot change into any one of them – indeed it does not undergo substantial change at all), but it resembles fire in being swift-moving and luminous.

225. Once again, a reference to *GC* 1, 331a23.

226. There seems to be no need to follow Vitelli here in suspecting a considerable lacuna in the text, if one assumes that in line 25 an *ê* ('or') has fallen out by haplography (after *êi*).

227. The 'force of the void' is a non-Aristotelian concept, which Philoponus introduced in his *Corollary on Place* (569,18-572,6); it may go back to the notion of 'following on to what is being emptied' (*pros to kenoumenon akolouthia*) as used by the physician Erasistratus (third century BC), on which see Furley and Wildberg (1991) 30 n. 24. It lived on in the medieval concept of *horror vacui*, on which see Grant (1981) 67-100. The void, in Philoponus' view, has no real 'force', but nature is such as to make sure that the substances of the world always form a continuum, so that the underlying three-dimensional spatial extension, which in its own nature is void, is never actually left empty. By referring to this principle here, Philoponus is definitely going beyond a mere exegesis of Aristotle's text.

228. *GC* 1, 318b27ff.

229. i.e. the objects in the sublunary world. The anonymous 'some' whose exegesis is here represented take the passage in Aristotle to be about the place of the whole cosmos, and they take Aristotle's view to be that the cosmos is in place in virtue of its (cosmic) parts, which are all ultimately located within the inner surface of the outer sphere.

230. The parts, being continuous, do not move *independently* of each other.

231. For these two alternative interpretations of what is meant by 'the place of the parts of the outer sphere', see above, 594,11-28 with nn. 206 and 207. On both interpretations, we are dealing only with a place 'by analogy' according to Philoponus, because the place at issue does not fulfil all formal requirements of an Aristotelian place (on which see above, n. 207): the convex surface of the sphere of Saturn does not fulfil the requirement of containing the emplaced body on all sides, the parts of the outer sphere themselves do not fulfil the requirement of immobility (nor, presumably, the requirement of being separable and, hence, independent), although the surface does serve to 'mark off' the object at issue. Although Philoponus rejects both solutions in his *Corollary on Place* (on which see above, n. 207), he appears to find the second one at least more congenial (he speaks of 'more properly' (*kuriôteron*)) within the context of Aristotle's account, possibly because it at least provides the outer sphere with an independent external surface.

232. Strictly speaking, they contain each other as parts of a *sphere*, yet they move in a circle, and the parts that exchange place accordingly do so in circles.

233. So also when we are dealing with the place of the outer sphere *as a whole*, we are faced with the same two interpretations we encountered a few lines ago when we were considering the place of the *parts* of the outer sphere.

234. For Philoponus' text, as indeed for Aristotle's text itself, the MSS vary between *eph' ho* and *eph' hôi*. (For Simplicius, see Diels *ad* 591,34.) Ross, while conceding that 'the evidence of the MSS and commentators is rather in favour of *eph' ho*', prefers *eph' hôi* on the strength of the parallel with *epi kuklôi* in line 13. His paraphrase is 'The direction in which the *ouranos* moves ... contains all the parts of the universe and is thus their place'; cf. Barnes' translation 'The line on which it is moved provides a place for its parts', and Hussey's 'but on the path they move along, in this way they are a place for their parts'.

235. *Phys.* 5, 227a6: 'A thing is consecutive (*ekhomenon*) if it is both successive (*ephexês*) and in contact (*haptêtai*)'.

236. 'Incidentally' in the broad sense explained above (partly with the help of Themistius) at 595,18-596,8, which includes 'in respect of something else'.

237. This takes up the two interpretations of 'parts' referred to above (600,26-8) in the comments on 212a35: we are speaking either of the heavens as a whole (in which case the parts at issue are the separate spheres), or of the outer sphere alone (in which case the parts at issue are the continuous parts of that sphere). In the former case, one part (the outer sphere) is not in a place in the proper sense. In the latter case no part is. See also above, nn. 206 and 207.

238. Most MSS of Aristotle omit 'only' (*monon*), as does Ross' edition; the confusion may be due to uncertainty about the referent of 'it': the universe or the heavens.

239. On this use of *ouranos* ('the heavens') as equivalent to 'the universe' (*to pan*), see above, n. 200.

240. For 'the universe' refers to everything there is; see above, n. 200.

241. Simplicius *in Phys.* 593,38-594,6 lists no less than five different explanations of the occurrence of 'possibly' (*isôs*) here, which include the fact that Aristotle has not yet established that the world is one, or that there is no void outside it (the subjects of *Cael.* 1.8 and 9).

242. The words 'immobile limit' are bracketed in the edition of Ross, who regards them as a later addition; they are absent from the texts of Themistius and Simplicius. Without them, Aristotle's text would read: 'And the place is not the heavens, but the inner part of the heavens that is in contact with the body that is capable of locomotion.'

243. See Philoponus' text above, 594,5-7.

244. See above, 504,10-514,6.

245. *Phys.* 4, 212b35-213a1: '[...] and what is in place is like a detached part in relation to the whole, as when one produces change in a portion of water or air'.

Select Bibliography

Algra, K.A. (1995) *Concepts of Space in Greek Thought*, Leiden/New York/Köln (Brill).

Blumenthal, H.J. (1996) *Aristotle and Neoplatonism in Late Antiquity: Interpretations of the* De Anima, London (Duckworth).

De Haas, F.A.J. (1997) *John Philoponus' New Definition of Prime Matter*, Leiden/New York/Köln (Brill).

Furley, D. and Wildberg, C. (1991) *Philoponus: Corollaries on Place and Void with Simplicius: Against Philoponus on the Eternity of the World*, tr. David Furley and Christian Wildberg, London (Duckworth).

Golitsis, P. (2008) *Les commentaires de Simplicius et de Jean Philopon à la* Physique *d'Aristote: Tradition et Innovation*, Berlin/New York (de Gruyter).

Grant, E. (1981) *Much Ado about Nothing: Theories of Space and Vacuum from the Middle Ages to the Scientific Revolution*, Cambridge (CUP).

Hoffmann, P. (1987) 'Sur quelques aspects de la polémique de Simplicius contre Jean Philopon: de l'invective à la réaffirmation de la transcendence du ciel', in I. Hadot (ed.) *Simplicius: sa vie, son oeuvre, sa survie*, Actes du Colloque International de Paris (28 sept.-1 oct. 1985), Berlin (de Gruyter), 183-224; English tr. in Sorabji (1987) 57-84.

Hussey, E. (1983) *Aristotle's* Physics *Books III and IV*, tr. with notes, Oxford (Clarendon).

Machamer, P. (1978) 'Aristotle on Natural Place and Natural Motion', *Isis* (69) 377-87.

Morison, B. (2002) *On Location: Aristotle's Concept of Place*, Oxford (OUP).

Ross, W.D. (1936) *Aristotle's Physics: A Revised Text with Introduction and Commentary*, Oxford (Clarendon).

Sedley, D.N. (1987) 'Philoponus' Conception of Space', in Sorabji (1987) 140-54.

Sorabji, R. (1983) *Time, Creation and the Continuum: Theories in Antiquity and the Early Middle Ages*, London (Duckworth).

Sorabji, R. (ed.) (1987) *Philoponus and the Rejection of Aristotelian Science*, London (Duckworth).

Sorabji, R. (1988) *Matter, Space and Motion: Theories in Antiquity and their Sequel*, London (Duckworth).

Sorabji, R. (ed.) (1990) *Aristotle Transformed: The Ancient Commentators and their Influence*, London (Duckworth).

Sorabji, R. (2004) *The Philosophy of the Commentators 200-600 AD, A Sourcebook*, vol. 2: *Physics*, London (Duckworth).

Todd, R. (2003) *Themistius: On Aristotle Physics 4*, tr. Robert B. Todd, London (Duckworth).

Urmson, J.O. (1992a) *Simplicius: Corollaries on Place and Time*, tr. J.O. Urmson, annotated by L. Siorvanes, London (Duckworth).

Urmson, J.O. (1992b) *Simplicius: On Aristotle Physics 4.1-5, 10-14*, tr. J.O. Urmson, London (Duckworth).

Verrycken, K. (1990a) 'The Development of Philoponus' Thought and its Chronology', in Sorabji (1990) 275-304.

Verrycken, K. (1990b) 'The Metaphysics of Ammonius, son of Hermeias', in Sorabji (1990) 33-274.

Waterfield, R. and Bostock, D. (1996) *Aristotle, Physics*, Translated with an Introduction and Notes, Oxford (Clarendon).

Westerink, L. G. (1962) *Anonymous Prolegomena to Platonic Philosophy*, Amsterdam (North-Holland Publishing Co.).

Wolff, M. (1987) 'Philoponus and the Rise of Preclassical Mechanics', in Sorabji (1987) 84-121.

English-Greek Glossary

able: *pephukos*
absolutely necessary: *pasa anankê*
absurd: *atopos*; *hupopsukhros*
accept: *prosiesthai*
accidental: *sumbebêkenai*; (*kata*) *sumbebêkos*
accommodate: *diasôizein*
account: *logos*; *apophainesthai*; *eirêke*
acquire form: *eidopoieisthai*
acquire volume: *onkousthai*
act: *poiein*
active: *poiêtikos*
actual: *energeiai* (*on*)
actual words: *lexis*
actuality: *energeia*; *entelekheia*
ad infinitum: *ep' apeiron*
add (to): *prostithenai*; *emballein*
addition, explanatory: *exêgêtikos*
adduce: *epagein*
admit: *dekhesthai*; *anadekhesthai*; *sunkhôrein*
admit of: *epidekhesthai*
affection: *pathos*
affirm: *phanai*
agree: *homologein*
air: *aêr*
analogously: *kata* (*tên autên*) *analogian*
analogy: *analogia*; by analogy: *kata* (*tên autên*) *analogian*
analyse: *dialambanein peri* + gen.
ancient (thinkers): *arkhaios*
angle: *gônia*; at equal angles: *pros isas gônias*; at right angles: *pros orthas gônias*
animal: *zôion*
announce beforehand: *proepaggellein*
antecedent: *hêgoumenon*
appear: *emphasis*; *phainesthai*
appearance: *phainomenon*; outer appearance: *morphê*
apply: *harmozein*; *huparkhein*
appropriate: *oikeios*
apt to receive: *dektikos*
argue for: *epikheirein eis*; argue on both sides *eis, eph' hekatera*
argument: *epikheirêma*; *epikheirêsis*; *logos*
assertion: *apophansis*
assimilate: *exomoioun*
assimilation: *proskrisis*
assume: *lambanein*; *prolambanein*
attempt to establish: *anaskeuê*
avoid: *mê*

base: *basis*
bear on: *pros*
belong to: *huparkhein*
bereft of: *esterêmenos* + gen.
between, in: *metaxu*
blossom up: *anablastanein*
body: *sôma*; *megethos*
body-like: *sômatikos*
bound: *horizein*
boundary: *horos*; *peras*; *perigraphê*
briefly: *suntomôs*
bring to light: *apodeiknunai*
bulk: *metron*

can: *dunasthai*; *endekhesthai*; can act on: *poiêtikos*; can be acted upon: *pathêtikos*
capable, naturally: *pephukos*
capable of being acted upon, not: *apathês*
capable of change: *kinêtos*; *pathêtikos*
capable of locomotion: *kinêtos*
capable of sense-perception: *aisthêtikos*
capacity: *dunamis*; defining capacity: *perioristikos*; limiting capacity: *peratôtikos*
cases, sampling: *epagôgikos*
categorical: *katêgorikos*
cause: *aitia*; *aition*; produce by a cause: *aitian*
centre: *kentron*; *meson*
change, to change: *kinein*, *kinêsis*; *metaballein*, *metabolê*; *exallattein*; *ameibein*; *alloiôsis*; without change: *ametablêtos*
change into: *metaballein eis*
change place: *metaballein* (*kata topon*)
change position: *metakinein*; *methistasthai*; change of position: *metastasis*
changing: *en kinêsei*
characterise: *kharaktêrizein*

characteristic: *huparkhon*
charitable: *eugnomôn*
circular (motion): *kuklôi*
circumference: *perimetros*
circumscribe: *perigraphein*
claim: *phanai*; *eipein*; *legein*; *nomizein*
clarify: *saphêneia*; *saphês*
clear: *phaneros*
cogent: *anankaion*
cognate: *sumphutos*
coincide: *epharmozein*; *en tôi autôi*; *kata tauton*; be made to coincide: *episuntithenai*
colour: *khrôma*
combination: *sunodos*; *sunthesis*; (*to ex*) *amphoin*; *sunamphoteron*; *holon*
come in together: *suneisienai*
coming-to-be: *genesis*; *genêtos*
common: *koinos*; common speech: *sunêtheia*
community: *koinônia*
compare: *exomoioun*
compelling: *anankaios*
complete: *entelês*; *sumplêrôtikos*
components: *suntithenta*
composed, be: *sunkeisthai*
composite: *sunthetos*; *sunêmmenon*
compound: *sunthetos*; *sunamphoteron*
compression: *pilêsis*
comprising: *holotês*
concave: *koilos*
conceive (of): *epinoein*
concentrate: *gumnazein*
concept: *ennoia*
conception: *ennoia*; *huponoia*; *prolambanein*
conclude: *manthanein*; *sunagein*
conclusion: *sumperainesthai*; *sumperasma*
condensation: *puknôsis*
condition: *skhesis*
confine: *horizein*
confines: *perigraphê*
conformity, be in conformity with: *analogein* + dat.
conjecture: *stokhasmos*
connect: *sumplekein*; *akolouthia*
connection: *sunaphê*
consecutive: *ekhomenos*
consequent: *hepomenon*
consider: *theôrein*
consist of: *einai ek*
constitute: *apotelein*; *poiein*; {*einai*}
construct argument: *kataskeuazein*
construct syllogism: *sullogizesthai*
construction, constructive overview: *kataskeuê*
contact, in: *haptesthai*; *ephaptesthai*; *pros*
contain, be contained: *dekhesthai*
contained in: *khôrêthênai en*
container: *dekhesthai*
contemplation: *thea*
continuous: *sunekhês*
contours: *perigraphê*; limiting contours: *perata*
contradict: *makhesthai*
contrary: *enantios*; *antikeimenos*; to nature: *para phusin*; opposition between contraries: *enantiôsis*
contribute to: *sumballesthai pros* + acc.; *suntelein eis*
conversely: *empalin*
conversion, convert: *antistrophê*
convex: *kurtos*
convincing, in a convincing way: *meta paramuthias*; without a convincing argument: *ektos paramuthias*
cosmic: *kosmikos*
cosmos: *kosmos*
count: *arithmein*
counterpart, state as the: *antapodidonai*
cover: *epekhein*; *epilambanein*
create: *dêmiourgein*
criterion: *kanôn*
criticise: *enkalein*
cut off: *apotemnein*
cut up: *katatemnein*

decrease: *meiôsis*; *phthisis*
decrease, to: *phthinein*
deem, be deemed to: *dokein*
defence: *apologeisthai*; *apologia*
define: *horizein*; *horizesthai*; *periorizein*
defining capacity: *perioristikos*
definite: *hôrismenos*
definition: *horismos*; *horos*; *logos*
definitional: *horistikos*
delimit: *peratoun*
delineate: *kataskeuazein*
demand: *apaitein*
demiurgic: *dêmiourgikos*
demolish: *anairein*
denominations, use different: *kalein diaphorôs*
describe: *legein*
destroy: *anairein*
detail: *akribês*
determine completely: *sumplêrôtikos*
difference, specific: *diaphora*
different: *diaphoros*; *diapherein*
differentia, differentiation: *diaphora*
difficult: *duskherês*

difficulty: *duskolia*; *khalepotês*
digress(ion): *parekbainein*
dimension: *diastêma*; *diastasis*; what has, is extended in, dimensions: *diastatos*
dimensionless: *adiastatos*
directly: *aparallaktôs*
disappear: *sunanaireisthai*
discordant: *plêmmelês*
discovery: *heuresis*
discrete: *diêirêmena*
discuss *episkeptesthai*; *legein*
discuss difficulties of: *aporein peri* + gen.
disorderly: *ataktos*
disprove: *elenkhein*
distended: *onkôtheis*
distinct: *diairetos*; *diêirêmenon*
distinction: *diairesis*; *diakrinein*, *diakrisis*; make relevant distinctions: *diorizesthai*
distinctive property: *idion*; *idiotês*
distinguish: *diakrinein*; *diairein*
divide: *diairein*
divisible: *diairetos*
division: *diairesis*; *tomê*
doctrine: *dogma*
duplicate, to: *diplasiazein*

early (thinkers): *palaios*
earth: *gê*
effect, have effect on: *suntelein*
elaborate: *gumnazein*
elegantly: *glaphurôs*
element: *stoikheion*
emplaced: *en topôi*
empty: *kenos*
enclose: *perigraphein*
encompass: *periekhein*; *emperilambanein*
end: *telos*
enquiry, make: *zêtein*
ensouled: *empsukhoumenos*
entirety, in its: *holos*
entity, whole: *holos*; separate: *khôristê*; discrete: *diêirêmena*
enumerate: *aparithmein*
equal: *isos*
equivalent: *anti*
escape notice: *kleptesthai*
essence: *ousia*; *to ti esti*
establish: *kataskeuazein*; *anaskeuazein*; *hupotithesthai*; *apodidonai*
establish proof: *deiknunai*
eternal: *aidios*
evident: *prodêlos*
exact: *akribês*
exactly: *akribôs*
examine: *elenkhein*; *apoblepein eis*
examination: *zêtein*; *theôria*
example: *paradeigma*; *epagôgê*; *epaktikôs*
exceed: *peritteuein*
exchange: *metaballein*; *ameibein*
exegesis: *exêgêsis*
exhaust: *katadapanan*
exhibit change: *kineisthai*
exhibit locomotion: *kineisthai kata topon*
exhibit rotation: see *kuklos*
exist: *einai*; *huphestêkenai*; *en hupostasei einai*; *huparkhein*; thing(s) that exist: *on*(*ta*); exist prior to: *proüparkhein*
existence: *einai*; way of: *huparxis*
existent: *en huparxei*; *huphestêkos*; naturally existent: *pephukôs*
existing thing: *on*
explain: *exêgeisthai*; *hermêneuein*; has been explained: *eirêtai*
explanation: *exêgêsis*; *aition*; *logos*
explanatory: *exêgêtikos*
express: *sêmainein*
extended, be: *diistasthai*; (*trikhêi* three-dimensionally) *diestêkenai*
extension: *diastêma*
external: *exôthen*
extreme: *eskhatos*

facing: *pros*
fact: *energeiai*; based on fact: *pragmateiôdês*
fall short: *elleipein*; *endein*
feature: *ti*; not a feature: *mêden*
feed: *trephein*
fiery region: *hupekkauma*
figure: *skhêma*
fill (up): *plêroun*; *eksplêroun*
filled: *plêrês*
final: *telikos*
find: *heuriskein*
fine structure: *leptomereia*
fire: *pur*
fit (onto): *harmozein*; *epharmozein*
fitting: *prosekhôs*
fittingly: *eikotôs*
fixed: *hôrismenos*; (sphere) of the fixed stars: *aplanês*
focus: *skopein*; *aphoran*; *gumnazein*
foetus: *kuêma*
follow: *hepesthai*; *akolouthein*; *parakolouthein*; *sumbainein*
followers: *hoi peri* + acc.

force: *bia*
form: *eidos*; *morphê*; *skhêma*;
possessing form: *eskhêmatismenos*
formally: *eidei*
formless: *askhêmatistos*
formulate: *apodidonai*; *ektithesthai*; *eipein*
fundamental: *kurios*; *arkhê*
fuse: *sumphuesthai*

general: *katholikos*; *holikos*; in general: *holôs*; *haplôs*
generate: *gennan*
generic: *genikos*
genus: *genos*
give: *apodidonai*
give way to: *parakhôrein*
go out together: *sunexienai*
grant: *didonai*; *sunkhôrein*
grasp: *lambanein*
great-and-small: *mega kai to mikron, to*
grow: *auxanein*; grow along with: *sunauxanein*

habit: *ethos*
happen: *gignesthai*
heaven(s): *ouranos*
heavenly: *ouranios*
height: *bathos*
hollow: *koilos*
homoiomerous: *homoiomerês*
homonymous: *homônumos*
hypothesis: *hupothesis*
hypothetical: *hupothetikos*

idea: *huponoein*
identify: *legein*
imagine: *phantazesthai*
immanent (form): *en autêi*
immediately: *prosekhôs*; *euthus*;
surrounding: *perix*
immobile: *akinêtos*; *ametakinêtos*; *êremein*
impede: *eirgein*
impeded, not: *akôlutôs*
implant: *endidonai*
imply: *paremphainesthai*
impossible: *adunatos*; make: *sunanairein*
improper(ly): *katakhrên*, *katakhrêstikos*
impulse: *hormê*
in between: *metaxu*
incidentally: *kata sumbebêkos*
include: *sumperilambanein*
incorporeal: *asômatos*
increase: *auxêsis*
increase, to: *auxanein*; *pleiô poiein*
indefinite: *aoristos*
indicate: *deiknunai*
indifferent: *adiaphoros*
individual: *heis*; *hekastos*; *meros*
individually: *idiai*
induction: *epagôgê*
inductive(ly): *epagôgikos*, *-ôs*; *epaktikos*, *-ôs*
inexplicable: *anermêneutos*
inference: *sunêmmenon*
infinite: *apeiros*
infinitely, ad infinitum: *ep' apeiron*
inhere: *einai*
inherent properties: *enuparkhonta*
inner limit: *eskhaton peras*
inseparable: *akhôristos*
instantaneous nature: *oxutês*
intellect: *nous*
intellection, object of: *noêma*
intelligible: *noêtos*
interchange, being interchanged: *exallagê*
interchangeable, use interchangeably: *epallattein*
internal: *en bathei*
interpenetrate: *khôrein dia*
intrinsic properties: *huparkhonta kath' hauto*
introduce: *eisagein*
introduction, theoretical: *theôria*
investigate: *skopein*; *theôrein*; *zêtein*
investigation: *skemma*; *skepsis*; *theôria*; *zêtêsis*
involve, be involved in: *paremphainesthai*

judgment: *doxa*
juxtaposed: *parallêlos*

keen: *akros*
keep together: *summenein*; *sunekhein*
kind: *eidos*

label, to: *onomazein*; *kalein*
language, everyday: *sunêtheia*
large, be too: *pleonazein*
leave: *perileipein*; *metaballein apo*
leave behind: *apoleipein*
lecture: *lexis*; *sunousia*
length: *mêkos*
let loose: *aphienai*
light, bring to: *apodeiknunai*
limit: *peras*; *eskhatos*; *periekhein*;
limit, to: *peratoun*
limited: *peperasmenon*
limiting capacity: *peratôtikos*
limiting contours: *perata*

limitless: *aperatôtos*
line: *grammê*
link, be linked with: *anienai*
list: *aparithmein*; *ektithenai*
located: *en*
locomotion: *phora*; *kinêsis kata topon*; be in: *kata ton topon kinoumena*
loose, *see* let loose
lose: *existasthai*

magnitude: *megethos*
maintain: *legein*
major: *meizôn*
man: *anthrôpos*
manifold variety: *diaphora*
mark off: *aphoristikos*; *aphorizein*
mass: *onkos*; *holos*
mathematical: *mathêmatikos*
matter: *hulê*
mean: *sêmainein*
meaning: *sêmainein*; *ennoia*; *nous*
measure: *metron*
medium: *organon*
mention: *mimnêiskein*
method: *methodos*; *tropos*, of proof *apodeiktikos*
minor: *elattôn*
mistake: *apatê*
mode, second: *deuteros*
moon, of the: *selêniakos*
mother: *mêtêr*
motion, movement: *kinêsis*; *kinein*
move: *kinêsin poieisthai*; *pheresthai*; *metaballein*; *metapherein*; *antimethistasthai*
move along: *summetapheresthai*; *summethistasthai*
move away: *methistasthai*
move downward: *katapheresthai*
mutual replacement: *antimetastasis*; *antiperistasis*

name: *onomazein*; *kalein*; *legein*
name after: *paronomazein*; *legein*
name-label: *prosêgoria*
natural: *phusikos*
naturally: *eikotôs*; *autophuôs*
naturally capable: *pephukos*
naturally existent: *pephukôs*
nature: *phusis*; *ousia*; contrary to: *para phusin*; student of: *phusikos*; instantaneous nature: *oxutês*
necessary: *anankaios*
necessity: *anankê*
negation: *antithesis*
neighbours; *perix*
next: *ephexês*
non-proper: in a non-proper way: *ou kuriôs*; use word in a non-proper sense: *katakhrên*
non-sensible: *ouk aisthêtos*
not-being: *mê on*
note: *ephistasthai*
notice *see* escape
notion: *ennoia*; *huponoein*
number: *arithmos*
numerically: *arithmôi*
nurse: *tithênê*

object: *pragma*; the same: *to auto*; mathematical: *mathêmatikos*; of intellection: *noêma*
objective: *skopos*
obscure, to: *epitholoun*
obtain: *katalambanein*; *tunkhanein*
obvious: *phaneros*
obviously: *prodêlôs*
occupy: *epekhein*; *katekhein*; *katalambanein*; *metalambanein*; *metempiptein*
offer: *apodidonai*
omit: *paralimpanein*; *parienai*
opinion: *doxa*; based on (received) opinions: *endoxos*
oppose: *antitattein*
opposed: *antikeimenos*
opposite: *enantios*; *antikeimenos*; be opposite, lie opposite: *antikeisthai*
opposition: *antithesis*; between contraries: *enantiôsis*
outer appearance: *morphê*
outline: *skhêma*; *malista*
overview, constructive: *kataskeuê*
own, in their own right: *kath' hauta*

part: *morion*; *meros*; in parts: *memerismenôs*; as a part: *merikos*
participant: *metalêptikos*; *methektikos*
participate: *metalambanein*
pattern: *logos*
perish: *apollusthai*; *phtheiresthai*
perishing: *phthora*; *phthartos*
perspective: *skhesis*
pervade: *khôrein dia* + gen.
physical: *phusikos*
place: *topos*
place-body: *topikon sôma*
place-point: *topikê stigmê*
plain to see: *prophanês*
plane: *epipedon*
plausible: *pithanos*
pleonastically: *parallêlos*
plural: *pleiô*
plurality: *plêthos*

point: *sêmeion*; *stigmê*
points at which heavenly bodies rise: *anatolika*
portable: *metaphorêtos*
portion: *metron*; *morion*
position: *thesis*; see also: change
possessing qualities: *pepoiômenos*
possibly: *isôs*
potentiality: *dunamis*
potentially: *dunamis*
power: *dunamis*
preconceived: *proeilêmmenos*
predicate: *katêgorein*
predominance: *epikratein*
pre-exist: *proüparkhein*
preferred: *proêgoumenôs*
preliminary discussion: *proêporêmenon*
premiss: *protasis*
prerequisite: *huparkhon*
present, be: *huparkhein*
present, to: *tithenai*; *protithenai*; *ektithesthai*; *paristanai*; *propherein*
preserve: *phulattein*; *sôizein*
presupposition: *prolambanein*; *hupolêpsis*; *axiôma*
prevent: *kôluein*
primary, primarily: *prôtos*; *malista*
prime (matter): *prôtê* (*hulê*)
principal: *kurios*
principle: *arkhê*; *stoikheion*; self-evident principle: *axiôma*
prior, be prior to: *proüparkhein*
probably: *isôs*
problem: *problêma*
process of growth: *auxêsis*
process of rotation: *kuklôi kinêsis*
produce by a cause *see* cause
productive: *poiêtikos*
productive of change: *kinêtikos*
profound: *megas*
pronounce: *eipon*
pronounce on: *apophainesthai peri*
proof: *apodeixis*; method of proof: *apodeiktikos*; set up proof: *anaskeuazein*; establish proof: *deiknunai*
proper: *oikeios*; *heautou*, *heautôn*; *idios*; be proper to: *huparkhein*; in the proper sense: *kurios*
properties: *huparkhonta*
property: *pathos*; be a property of: *huparkhein*; *sumbebêkenai*; distinctive: *idion*; *idiotês* ('property' has no counterpart in the Greek at 605,5)
proponent: *proïstasthai*
prove: *apodeiknunai*
provide: *apodidonai*; *antapodosis*
provide room: *khôrein*
provide place: *einai*
psychic: *psukhikos*
push away: *ôthein*
put forward: *apophainesthai*
put together: *suntithenai*
puzzle, puzzlement: *aporein*; *aporia*
puzzling: *aporos*

qualification, without (further): *haplôs*
quality: *poiotês*; possessing qualities: *pepoiômenos*; without quality: *apoios*
quantify: *posoun*
quantity: *posos*
question at stake: *prokeimenon*

raise a difficulty: *aporein*
raise a puzzle: *pherein aporian*
rank, to: *tattein*
rarefaction: *manôsis*
rational: *logikos*
reading: *graphê*
reality: *ta onta*; *energeia*
reason: *aitia*; it stands to, is reasonable, reasonably: *eulogos*; *eikotôs*
reasoning: *logos*
recede: *hupeikein*; *hupokhôrein*
receive: *dekhesthai*, be *dektikos* of; *anadekhesthai*
receptacle: *dexamenê*; *dokheion*
reduce: *anagein*
refute: *elenkhein*
regain: *katalambanein*
region, fiery: *hupekkauma*
relate: *ekhein*
relation: *skhesis*
relative: *skhesei*; *thesei*; things that are relative: *pros ti*
relative, position: *thesis*
relative to: *pros*; things that are relative to: *pros ti*
relatively: *skhesei*; *thesei*
relevant: *sumballesthai*; *diorizesthai*
remain: *menein*
remind: *hupomimnêiskein*
remove: *aphairein*; *anairein*; *sunanairein*; *metakinein*
repeat: *analambanein*
replace: *antimethistasthai*; *metempiptein*
replacement: *metastasis*; mutual: *antimetastasis*
require: *apaitein*; *axioun*
resemblance: *homoiotês*
respect, in respect of: *kata*, in respect

of itself: *kath' hauto*; in respect of something else: *kat' allo*
rest: *êremia*; be at rest: *êremein*
result: *sumbainein*
revolution: *periphora*; *kuklôi phora*; *kuklos*
riddling: *ainittesthai*
ridiculous: *geloios*
right, in their own: *kath' hauta*
rightly, more: *mallon*
room: *khôra*; provide room: *khôrein*
rotate: (*kuklôi*) *kineisthai*
run away: *diarrhein*

safe: *asphalês*
safeguard: *phulattein*
same: *to auto*
sampling: *epagôgikos*
save: *sôizein*; *phulattein*
see (to be): *theôrein*
see (to be) present in: *entheôrein*
seed: *sperma*
seek: *ephiesthai*; *zêtein*
seeking, being sought: *ephesis*
seem: *dokein*
self-evident (principle): *axiôma*
self-mover, -moving: *autokinêtos*
sense: *sêmainomenon*; in what sense: *pôs*; in a general sense: *holôs*; in a common sense: *koinos*; in a sense: *pôs*
sense-perception: *aisthêsis*; capable of sense-perception: *aisthêtikos*
sensible: *aisthêtos*
separable, separate: *khôristos*; *diairetos*; *heteros*
separate, to: *khôrizein*; separate from, be separated from: *ekpiptein ek*; separate off: *aphorizein*
separately: *idiai*; *kekhôrismenôs*
separating: *khôrizein*
set in motion: *kinein*
set off: *apokrinein*
set out: *diastellesthai*; *ektithesthai*
shake: *saleuein*
shape: *skhêma*
share: *koinônein*; have a share in: *epikoinônein*
show: *deiknunai*; show not to exist: *anairein*
side: *pleura*; argue on both sides, *see* argue
sideways: *enkarsiôs*
simple: *rhaidios*; *eukherês*
simpliciter: *haplôs*
singular: *haplôs*
size, of equal: *isoonkos*
solution: *lusis*; solutions, discussion with: *proêuporêmenon*
solve: *luein*; *epiluesthai*
soul: *psukhê*
sound: *hugiês*
space: *khôra*; *topos*
species: *eidos*; *merikos*
specific: *merikos*
specify: *aphorizein*
speech, common: *sunêtheia*
sphere: *sphaira*
spherical: *sphairikos*
start: *arkhê*
state: *hexis*; *ekhein*
static: *menein*; *êremein*
stationary condition: *skhesis*
stick to (conception): *phulattein*
stop: *histasthai*; decide to stop: *apoklêrôsis*
straight, in straight lines: *ep' eutheias*; *kat' eutheian*; *kat' euthuôrian*
strange: *atopos*
strive: *ephiesthai*
striving, object of: *ephetos*; *orektos*
structure, fine: *leptomereia*
student of nature: *phusikos*
subject: *hupokeimenos*; *pragma*
subject to coming-to-be: *genêtos*
subject to perishing: *phthartos*
subject-matter: *prokeimenon*
subsist, have come to: *hupostênai*; subsist before: *proüpostênai*
subsistent thing: *hupostasis*
substance: *ousia*; a single: *monoeidês*; of wine: *nastotês*
substantial: *ousiôdês*
substrate, as a: *hupokeimenos*
subtract: *aphairein*
successive: *ekhesthai*; *ephexês*
suggestion: *huponoia*; produce stock of suggestions: *proêuporein*
suitable: *prosphuês*
summit: *akros*
supervene: *epigignesthai*
support: *kataskeuazein*; *paristanai*
suppose: *hupolambanein*; *tithenai*
supposed, be: *nomizesthai*; *prokeisthai*
supposition: *hupolêpsis*
surface: *epiphaneia*
surprising: *paradoxos*
surround: *periekhein*; *peri*; *perix*; surroundings: *periekhon*
suspended: *kremamenos*
sustain: *anekhein*; *sunistanai*
swap: *exallattein*
switch: *enallattein*
syllable: *sullabê*

syllogism: *sullogismos*; form a: *sullogizesthai*
syntactic(ally): *suntaxis*

take: *lambanein*; to be: *eklambanein*
take away: *aphairein*
take on: *proslambanein*
take up: *analambanein*; *epilambanein*; *katalambanein*
tally: *sumbolon*
teachings: *sunousia*
tend to: *thelein*
tendency: *rhopê*
term: *horos*; *onoma*
testimony: *marturia*
that in which: *to en hôi*
theorem: *theôrêma*
theoretical introduction: *theôria*
theoretical understanding: *theôria*
theory: *logos*; *theôria*
thing: *pragma*
thing(s) that exist(s): *on(ta)*
think: *nomizein*; *noein*; *huponoein*
this here, a: *tode ti*
thought: *epinoia; dianoia*
three-dimensionally extended, be: *trikhêi diestêkenai*
time: *khronos*
time-honoured: *palaios*
together: *hama*
top: *koruphê*
totality: *holotês*; in its own totality: *holon kath' holon heauto*
touch: *haptesthai*
tout court: *haplôs*
transcendent: *exêirêmenos*
transfer: *metaballein*
transform: *skhêmatizesthai*
transport: *metapherein*
travel along: *summetabainein*
triangle: *trigônon*
true: *alêthês*; be true of: *huparkhein*

ultimate: *eskhatos*
unaffected: *apathês*
unclarity: *asapheia*
unclear: *asaphês*
unclearly: *asaphôs*; *amudrôs*
undergo: *hupomeinein*
underlying: *hupokeimenos*
understanding: *katalêpsis*; *gnôsis*; *nous*; *theôria*
unify: *sunenoun*
universe: *to pan*
unqualifiedly: *haplôs*
unrelated: *askhetos*
unsubtle: *akompsos*
unwritten: *agraphos*
usage, use: *khrêsis*

vanish: *phtheiresthai*
variety (manifold): *diaphora*
verging on: *pros* + dat.
vessel: *aggeion*
view: *doxa*
virtue, in virtue of: *see* respect, in virtue of
void: *kenon*
volume: *megethos*; acquire: *onkousthai*

walking upright: *orthoperipatêtikos*
wall (of a pit): *plagioi*
water: *hudôr*
way: *tropos*
way of existence: *huparxis*
weight: *baros*
well-argued, in a well-argued way: *meta kataskeuês*
what it is: *to ti esti*
what it was (for this thing) to be: *to ti ên einai*
whole, as a: *holos*; *katholou*; *(ha)pas*
wholeness: *holotês*
wider: *epi pleon*
width: *platos*
without quality: *apoios*
witness: *theôrein*
wont, are wont to be: *thelein*
word: *phônê*; *lexis*
wording: *lexis*
words: *lexis*
work (written): *pragmateia*
world: *ouranos*

yield: *antiparakhôrein*

Greek-English Index

Abbreviations
Ar.: quotation from Aristotle
ass.: associated with (in the context)
def.: defined or specified, definition or specification
ex.: examples, exemplified
non-t.: non-technically
opp.: opposed to, opposite (in the context), in contrast to
Pl.: quotation from Plato
Them.: quotation from Themistius

Subject Index

Abbreviations
Ar.: Aristotle
def.: definition, defined
opp.: (as) opposed to
Philop.: Philoponus

Page numbers 1 to 12, divided by commas followed by spaces, refer to pages of this book. Numbers of the form 504,10.14; 510,16 refer to pages and lines of the Greek text as edited by Vitelli, which are given in the margin of the translation. Notes to the passages referred to are tacitly included.

CPSIA information can be obtained at www.ICGtesting.com
Printed in the USA
LVOW07s1314111015

457799LV00004B/113/P

9 781472 558008